Plate Tectonics for Introductory Geology

John R. Carpenter
University of South Carolina

Philip M. Astwood
Winthrop College

KENDALL/HUNT PUBLISHING COMPANY
Dubuque, Iowa

Copyright © 1983 by Kendall/Hunt Publishing Company

Library of Congress Catalog Card Number: 82–83989

ISBN 0–8403–2895–8

B   402895   01

Contents

Preface

This book evolved from our separate teaching experiences over the past several years. Independently, we arrived at a decision that Introductory Geology could be more than "learning about rocks." It is our contention that an introductory geology course could be centered on "plate tectonics," that marvelous theory introduced in the 1960's, that provides earth scientists with a hitherto unavailable model on which to explain most of the internal processes of the earth, and many of the surface processes as well. The plate tectonics concept is both simple and elegant, and easily grasped by students with little or no previous earth science background. We have found that by centering our courses around this concept, students gain a more comprehensive view of geology. It has been exciting to us to see our past students gain this more comprehensive view. Their biggest complaint—their textbook didn't follow our teaching pattern. Out of a desire to make our courses even better was born the concept of this book.

Any textbook owes much to many people. Certainly this book does. We owe a debt of gratitude to those earth scientists whose ideas shaped the concept of plate tectonics. We owe a lot to our past students who hassled us about their textbook and asked why we didn't do something about it. But mostly we owe a tremendous debt of gratitude to Ms. Nancy Siegel of Kendall/Hunt, who saw from an editor's eye the value of the approach we were taking. She both encouraged and cajoled us throughout to finish this task. It was she who inspired us to take that first crude draft off of the shelves (where it had been collecting dust for two years), to smooth it out and publish it.

Several other people have aided us greatly in the preparation of the manuscript. We especially appreciate the critical review of the early draft of the chapters on sedimentation by Dr. Michael A. Arthur, and the critical review of the early draft of the chapter on the evolution of the Appalachians by Dr. Robert D. Hatcher, Jr., both of the University of South Carolina. Pris Ridgell and Sara Strait typed what must have seemed to them endless revisions. Melinda Prim, an art student at the University of South Carolina, is responsible for most of the original illustrations.

Finally, we thank our wives, Charlie Carpenter and Valarie Astwood, for their technical assistance, but mostly for their constant support and encouragement. Without them, it couldn't have been done. To them we dedicate this book.

CHAPTER 1
Overview of Plate Tectonics

Have you ever marvelled that the western coastline of Africa and the eastern coastline of South America appear to be mirror images of each other? Probably you have been made aware of this resemblance in some phase of your school work, even if you have not made this discovery by yourself. Alfred Wegener, in the early part of the 20th century, made this discovery and attempted to account both for the similarity and for the distance separating these continental masses by a process he called **continental drift.**

Wegener's continental drift idea met with both acceptance and skepticism and has enjoyed a real rollercoaster ride in terms of acceptance. Perhaps the most unusual aspect of this concept is that so many trained scientists, ourselves included, were able to explain away the resemblance as "coincidental" and did not see the ramifications of Wegener's idea. Today, the vast majority of earth scientists concur that the continental masses are not in a permanent location, nor have they been in the past, nor will they be in the future.

The earth sciences have experienced a full-scale conceptual "revolution" in the past twenty years. No longer must we wonder why earthquakes and volcanoes recur so often in the same zones. No longer must we find it remarkable that folded mountain belts and continental margins are roughly parallel. No longer must we marvel that the Mid-Atlantic Ridge lies almost exactly halfway between North America and Europe, and South America and Africa. We now have a basic concept by which we can explain these as well as other previously baffling observations such as volcanic lava generation, island arcs, deep oceanic trenches, and long, straight continental block-faulted valleys. The concept on which we can base our explanation of the above phenomena is known as **plate tectonics,** and it is an outgrowth of Wegener's continental drift. This book will use plate tectonics as a central theme in discussing a series of processes by which we can explain most internal and some surface processes. Several other introductory geology texts have taken a related approach, but have generally attempted to pull together concepts with this basic model only after a study of the internal and surface processes. In this book, we begin with an overview of plate tectonics, then proceed to two chapters on the nature of the whole earth (including a very brief section on rocks and minerals) and geologic time.

These two chapters are necessary to provide you with some basic concepts and definitions. We return to a detailed discussion of the processes of rock genesis later, where we can talk of these processes in a plate tectonics framework.

Plates of the Earth

There is very little argument today among members of the geologic community that the upper layer of the earth has been and continues to be unstable. There is overwhelming evidence that sections of this uppermost layer are moving with respect to one another. "How" and "why," we may not know for sure, but we are certain that movement is taking place, and this movement gives rise to processes and products that can be observed on the surface. We will approach the concept of plate tectonics by looking first at the big picture, then examining the underlying evidence.

The upper layer of the earth (including the surface) can be looked at as a huge, spherical jig-saw puzzle, composed of pieces having length, width, and thickness. They are rather thin with respect to their length and width. Earth scientists called these pieces "plates," and the movement of these plates, absolutely and relatively to one another, **plate tectonics.** Tectonics is the study of the broader structural features of the earth (such as mountain building, volcanic activity and earthquakes) and their causes. Figure 1.1 is a world map with the plates named. To be sure, there are smaller plates and sub-plates that will have to be examined later, but for now let's concentrate on the larger, well-defined plates.

One feature of these plates needs to be pointed out immediately, because it gives rise to a fundamental difference between the concepts of continental drift and plate tectonics. Note that the plates are composed of both continental crust and oceanic crust. The old continental-drift concept considered movement only by the continental crust. In fact, this continental crust, having lower density than ocean floor crust, was thought to "float" across the deeper, more dense oceanic crust. Now it seems certain that continental and oceanic crust, together with some of the underlying mantle, are intimately associated in any given plate.

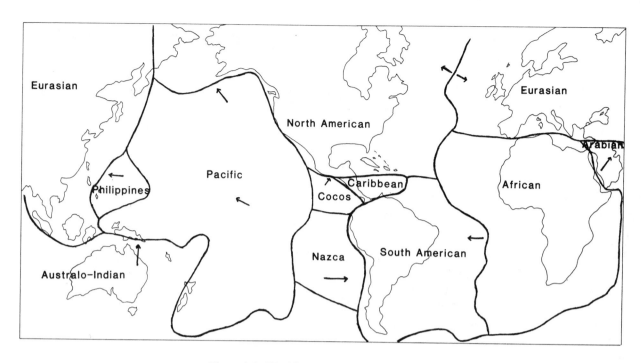

Figure 1.1. World map showing major plates.

Plate Movements

Plates are moving both absolutely and relatively to one another in response to very large forces in the mantle beneath the plates. We will see that it is important to understand both kinds of movement. First, however, let's consider relative movement by itself.

Relative movement simply means that plates are moving toward, away from, or parallel to each other (shown diagrammatically in Figure 1.2). When plates move relatively to one another, friction acts on the edges of plates, inhibiting the movement and storing up huge quantities of energy. When the forces that are pushing the plates overcome the friction along plate boundaries that are inhibiting movement, sudden movement takes place, and much energy is released in the form of earthquakes. The movement of plates causes the plates to heat up and, in some areas, causes the plate material to begin to melt, generating **magma,** a chemically complex rock melt.

Relative movement of plates parallel to one another, but in opposite directions, yields faults and earthquakes, but no magma is generated.

Relative movement of plates away from one another generates major fractures or openings in the plates perpendicular to the direction of relative plate movement. Magma generated at depth rises to the surface, creating new crust in the form of long, linear volcanic mountain chains. This can happen whether the plate boundary lies within oceanic crust or continental crust. One very obvious thing that takes place under these conditions is the generation of earthquakes when the surface rocks break apart. Second, as these cracks occur, the land *in between the cracks* can drop, form-

ing a long, low, linear valley. Examples of this kind of valley occur as the East African Rift Valley and the low valleylike part of undersea mountain chains like the Mid-Atlantic Ridge. Thus, the Mid-Atlantic Ridge is a volcanic mountain chain, split down the middle by the pull-apart forces.

Let's next consider what happens when these plates move toward one another. Obviously, unless the earth is getting larger (and it apparently is *not*), when plates move toward one another, a collision must take place. These collision boundaries, in many ways, are responsible for some of our most important physiographic features. Where collisions occur in which both leading plate edges are oceanic, deep curved or arcuate trenches occur, such as the Marianas Trench in the Pacific, the deepest known point on the surface of the earth. Likewise, when one leading edge is oceanic and the other continental, long, linear deep trenches form adjacent to the continental mass. An example of this kind of feature is the Peru-Chile Trench off the west coast of South America. In both cases, trenches form because one plate edge is forced *under* the other; in the case of ocean-continent collision, the oceanic edge is forced under the continental edge. In the case of ocean-ocean collision, *one* of the oceanic plates is forced under the other. (It doesn't seem to matter which one is forced under.) In all cases, the movement of plate edges toward one another generates earthquakes due to the violent nature of the release of stored-up energy, much as in the case explained earlier.

Recall that we said that when the plates move past each other, heat is generated. In the case of collision contact, the heat is sufficient to melt part of the forced-

2

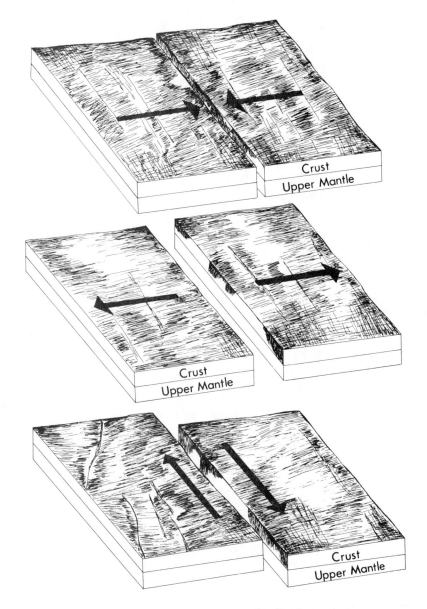

Figure 1.2. Three types of plate motions.

under plate, generating **volcanoes,** thus explaining the presence of such paired features as the Andes Mountains of South America with the Peru-Chile Trench and the arcuate Marianas Islands with the Marianas Trench. Both of these mountain chains are volcanic chains, unlike the Appalachians or the Himalayas.

Let's next consider what happens when both leading plate edges (moving on a collision course) are continental in nature. Apparently, because of the lightness (low density) of the material and the fact that continental crustal material is considerably thicker than oceanic crustal material, when there occurs a continent-continent collision, much buckling or **folding** of rocks occurs, along with heating up of the rock material. The end result is a very high chain of mountains in which rocks formed from sediments that once were deposited near or at shorelines have become changed

by the heat and pressure of the collision and are folded into long, high mountain chains. This is apparently the genesis of the Appalachian Mountains and the Himalayas, the home of the highest point on the earth's surface, Mt. Everest.

This summary concludes our introductory chapter on plate tectonics. The remainder of the book deals with the details of these processes and physiographic features and describes the origins and nature of the products of plate tectonics in a much more comprehensive way. Let us leave you with a chart which you will see again, upon which much of the rest of this book will be based. Let's look at the related *processes* and *physiographic features* that are generated by plate tectonics in Table 1.1. This chart will be reproduced, with products, toward the end of the book.

Table 1.1. Plate Motions and Physiographic Features

Process	Physiographic Features
Ocean—Ocean split	Volcanic mountain chain, e.g., Mid-Atlantic Ridge
Ocean—Ocean collision	Trench and volcanic mountain chain, e.g., Caribbean and Marianas Islands
Continent—Continent split	Long, linear faulted valleys, e.g., East African Rift Valley
Continent—Continent collision	Folded mountain chain, e.g., Appalachians and Himalayas
Continent—Ocean split	Occur only rarely
Continent—Ocean collision	Trench and volcanic mountain chain, e.g., Andes and Peru—Chile Trench
Parallel plate movement	Fault, e.g., San Andreas Fault

CHAPTER 2
Nature of the Planet Earth

Because the earth does not stand alone in the universe, we feel that it cannot be understood without understanding its relation to other parts of the solar system and universe. Therefore, we begin this book with some basic cosmic concepts (and definitions). Let's start by looking at the solar system and some of the physical characteristics of its planets. (Figure 2.1 and Table 2.1)

Solar System

The earth is one of the "minor" planets of the solar system, those small planets closest to the sun that have physical characteristics similar to one another and very different from the physical characteristics of the larger, farther out "major" planets. For those of your unfamiliar with scientific mathematical notation, we direct you to Table 2.2. It is imperative for your present and later understanding that you be familiar with this notation.

Our factual knowledge of the composition of the solar system is limited to:

1. spectral composition of the surface of the burning sun
2. physical properties of the various planetary bodies
3. composition of the outermost shell of the earth, the crust, and composition of meteorites and moonrocks

Table 2.2. Scientific Notation

$10^1 = 10$
$10^2 = 100$ (one hundred)
$10^3 = 1,000$ (one thousand)
$10^6 = 1,000,000$ (one million)
$10^9 = 1,000,000,000$ (one billion)

What can we then *speculate* about the origin of the sun and the solar system? Astronomers entertain a number of hypotheses or theories, too numerous to discuss here. Let us look at one hypothesis, recognizing at the outset that it has limitations and drawbacks but, for the most part, satisfies the curiosity of the authors of this book.

Let's examine the "nebular hypothesis," first proposed many years ago and still accepted by some astronomers. According to this hypothesis, the solar system, just prior to 4.6 billion years ago, was thought to have existed as a "nebula" or dust cloud. This nebula is thought to have originated from a **super-nova** explosion and consequent death of a pre-existing star. The nebula is thought to have begun contracting due to gravitational attraction between bits of matter. As this matter accumulated into a smaller, more dense and better defined central mass, the bits of matter fused together, creating new forms of matter and releasing a great deal of energy. This central mass became the star we know as our sun. While the bulk of the mass of the nebula was contracting to form the sun, it is thought

Table 2.1. Physical Characteristics of Our Sun and Planets in the Solar System

	Distance from Sun (x 10^6 km)	Radius (x10^3 km)	Mass (x10^{27} g)	Density (g/cm^3)
Sun		694	2×10^6	1.4
Mercury	58	2.4	0.4	5.5
Venus	108	6.1	4.9	5.3
Earth	150	6.4	6.0	5.5
Mars	228	3.4	0.6	3.9
Asteroid Belt	400	—	?	3.5
Jupiter	778	71.4	1.9×10^3	1.3
Saturn	1420	60.5	568	0.7
Uranus	2870	23.6	87.3	1.2
Neptune	4490	24.8	103	1.7
Pluto	5980	3.2	0.6	4

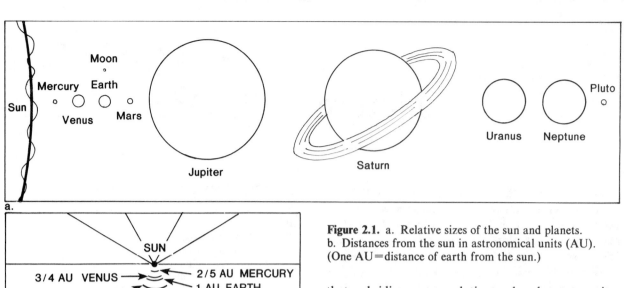

a.

SUN

3/4 AU VENUS → ← 2/5 AU MERCURY
1 1/2 AU MARS → ← 1 AU EARTH

← ASTEROID BELT

— 5 AU JUPITER

— 10 AU SATURN

— 20 AU URANUS

— 30 AU NEPTUNE

— 40 AU PLUTO

AU = ASTRONOMICAL UNITS

b.

Figure 2.1. a. Relative sizes of the sun and planets. b. Distances from the sun in astronomical units (AU). (One AU=distance of earth from the sun.)

that subsidiary accumulations, also due to gravitational attraction, led to the formation of the planets. So far, our hypothesis satisfactorily explains the position of the planetary bodies and the fact that the bulk of the mass of the solar system resides in the sun. One problem remains. Most of the **kinetic energy** (the energy of motion) of our solar system resides *not* in the sun, but in the planets. How was this kinetic energy transferred? We don't know for certain; this remains an unsolved mystery.

Early History of the Earth

The earth, like most of the other planets, is thought to have grown by **accretion,** the gradual collection of solar dust into a progressively larger mass. This accumulation of cosmic debris no doubt would have led to a chemically homogeneous planet. But we know that the earth is *not* chemically homogeneous. We will examine the evidence leading to this conclusion later. Right now, suffice it to say that geologists now are confident that the earth is quite *heterogeneous,* composed of an outer thin crust, a thick (2900 km) intermediate depth mantle, and a thick (3470 km) internal core. Figure 2.2 shows a cut-away picture of the interior of the earth.

If the earth began as a homogeneous mass and is now heterogeneous, we must somehow account for the distribution of material in the earth into a core and mantle, thought to be both physically and chemically dissimilar. Some process such as melting of the earth (either completely or at least partially) must be invoked to account for *re*distribution of materials within this relatively homogeneous mass.

To examine this possibility, we will now consider the *rate* of accretional growth of the earth. Did it occur slowly or rapidly? Slow accretion would have allowed the energy released by accretion to dissipate into space and would probably have resulted in a relatively cold early earth. Gradual release of heat energy from the breakdown of radioactive elements, those elements that

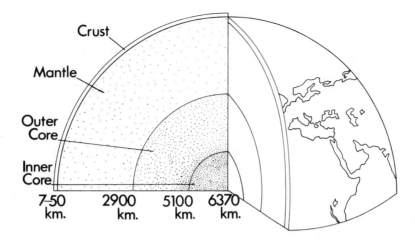

Figure 2.2. Internal structure of the earth.

spontaneously disintegrate, would have led to a slow warming of the earth, partial melting, and a gradual redistribution of matter into a core and mantle. Rapid accretion would not have allowed the energy released from gravitational "in-pull" to dissipate and could have led to an early "hot" earth, molten or partially molten, and a rapid redistribution of matter into a core and mantle. Whichever process may have occurred, the result is that some time fairly early in the history of the earth, the material constituting the whole earth separated into two basic shells, yielding a layered, nonhomogeneous earth composed of core and mantle. (At this point, you might be wondering, "What about the crust? Where did it come from? And how does it fit into this scheme of things?" The development of a crust is a later process, one which is still going on, the details of which will be addressed later.)

Composition of the Earth

Now, let's examine the earth in more detail. We *know* the mass, volume and density of the earth, and we know roughly the composition of the surface material. From this information, we can deduce certain qualities of the *interior* of the earth. The overall density of the earth ($\sim$ 5.5 g/cc) and the composition and density of surface and near-surface rocks ($<$ 3.0 g/cc) suggests that the earth is not homogeneous throughout. *This is a very important assertion and one on which will rest the theories about the processes by which the earth has evolved and the products of these processes.*

Another bit of evidence by which we assume a heterogeneous earth is based on **seismicity** (the study of the velocities and paths of deep earthquake waves). The evidence from seismic waves independently verifies a heterogeneous, *layered* earth composed of three main concentric shells:

1. the crust, a thin rocky veneer, composed of several different rock types of fairly low density (2.5–3.0 g/cc)
2. the mantle, a thick shell more homogeneous than the crust and composed of more dense rock (or a few rock types) with a density of 3.0–3.5 g/cc
3. the core, the central ball of the earth probably mostly metallic iron and nickel, generally thought to be liquid but perhaps solid in the very center.

We will examine in detail later in Chapter 5 some aspects of seismicity. Figure 2.3 shows how density, temperature and pressure vary with depth.

One other piece of evidence concerning the nature of the earth's interior comes from data on the composition of meteorites, those mysterious extraterrestrial bits of rock thought to be parts of the asteroid belt gathered when their path brings them close enough to the earth to be captured. Meteorites are important in two ways. First, they are thought to represent some of the oldest solar system material that has undergone little reworking. Second, because of their origin in the asteroid belt, they are thought perhaps to be similar in composition to the earth. Let's examine the composition of meteorites. In general, they can be categorized as follows:

1. "stones"—rock-like in composition, composed of several minerals
2. "irons"—metallic fragments composed mainly of iron and nickel
3. "stony-irons"—combinations of metallic and rock material.

Now, at first glance, that second possibility mentioned above, that meteorites have compositional similarities with the earth, seems remote, if not improbable, because meteorites are *not* similar to common rocks of the earth's crust. However, recall that the earth is believed to be compositionally layered, and the compo-

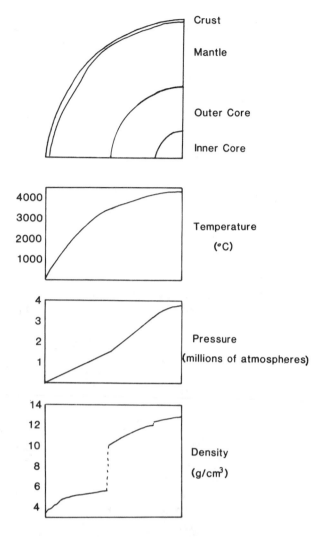

Figure 2.3. Variation of temperature, pressure and density through the earth.

1. Minerals are naturally occurring and inorganic.
2. Minerals are crystalline substances with a defined internal arrangement of atoms.
3. Minerals have chemical composition and physical properties which are either fixed or can vary within specific limits.

Let's examine each of these constraints in order.

First, minerals are naturally occurring and inorganic. The rule "naturally occurring" separates minerals from man-made crystalline substances (chemical solids) which might not occur under natural conditions. We run into some difficulty when we acknowledge the presence of man-made (synthetic) minerals such as synthetic diamonds, synthetic rubies, etc. Strictly speaking, the synthetic forms are *not* minerals even though they have the same composition and physical properties as the natural forms. By "inorganic," we mean that minerals are *not* a product of any life process. (Again, we run into some trouble with our definition when we consider the crystalline constituents of plants and animals, e.g., the calcite and aragonite of animal shells and the phosphate of bone material.)

By the term "crystalline substance," we mean that the atoms that combine to form minerals are arranged in a defined geometric pattern, with each mineral having its own unique atomic pattern. These atomic patterns govern specific physical properties. With this "rule," we eliminate from the family of minerals such things as naturally occurring glasses (e.g., volcanic glass), because by definition, a glass does not have an orderly geometric pattern of atoms. Rather, glasses are really supercooled liquids, with a chaotic arrangement of atoms. Finally, any mineral has a limited range of chemical composition and physical properties. Some minerals, such as quartz and pyrite, have essentially fixed chemical compositions. Quartz is silicon dioxide (SiO_2), and pyrite is iron disulfide (FeS_2). Other minerals, such as olivine, have variable chemical compositions. Olivine is a **silicate** mineral (containing silicon and oxygen) containing also either iron, magnesium or both. Chemically, the composition of olivine is written as: $(Fe,Mg)_2SiO_4$ with the parentheses indicating that it can contain either iron, magnesium or both. As the chemical composition of a mineral changes, so does its physical properties. (By deduction, you should be able to figure out that minerals with fixed chemical compositions tend to have fixed physical properties.) The physical properties most affected by chemical composition are:

1. **color.**
2. **hardness.**
3. **density** (mass/volume).
4. **cleavage** (ability to split along repeating planes of atoms).

sition of the "irons" coincides with the assumed composition of the earth's core, *and* the "stones" coincide with the assumed composition of the earth's mantle. Now before getting too astounded about these "coincidences," be aware that knowledge of the composition of meteorites was of prime importance in establishing a compositional model of the earth. The model was also subjected to other scrutiny and had to account for measured physical properties of the earth, such as density at high pressure and seismic-wave velocity propagation.

Minerals of the Earth's Crust

Minerals are the building blocks of rocks. Most of you are already familiar with some minerals—quartz from beach sands, mica, pyrite (fool's gold) and perhaps the highly magnetic mineral, magnetite. What exactly are minerals? In brief, minerals must adhere to the following "rules":

Take olivine, for example. The pure magnesium olivine is very light green in color and has a density of 3.27 g/cc; the pure iron olivine is very dark green (almost black) in color and has a density of 3.37 g/cc. Olivines with both iron and magnesium are an intermediate green in color and have a density intermediate between the values given above, depending on the exact proportion of iron to magnesium.

Minerals are classified on the basis of their chemical composition into the following groups:

1. **Silicates**—Containing silicon, oxygen and other elements.
2. **Oxides**—Containing oxygen and other elements.
3. **Carbonates**—Containing carbon, oxygen and other elements.
4. **Sulfides**—Containing sulfur and other elements.
5. **Sulfates**—Containing sulfur, oxygen and other elements.
6. **Native elements**—Containing only one element.

Silicates are quantitatively the most abundant mineral group in the earth's crust. Examples of silicates include:

1. **Feldspar**—A family of silicates all containing aluminum; individual members contain either calcium, sodium or potassium. Feldspars are divided into two sub-families:
   a. potassium feldspar—containing potassium, aluminum, silicon and oxygen ($KAlSi_3O_8$);
   b. plagioclase feldspar—contains calcium and/or sodium, aluminum, silicon and oxygen ($CaAl_2Si_2O_8$—$NaAlSi_3O_8$).
2. **Quartz**—SiO_2.
3. **Mica**—Another family of minerals all containing aluminum and hydrogen; sub-families contain iron and/or magnesium or potassium. Micas are divided into two sub-families:
   a. biotite—black mica, chemically $K(Fe,Mg)_3AlSi_3O_{10}(OH)_2$.
   b. muscovite—white mica, chemically $KAl_3Si_3O_{10}(OH)_2$.
4. **Olivine**—Containing iron and/or magnesium.
5. **Pyroxenes**—A chemically complex family of iron-magnesium silicates.
6. **Amphiboles**—Another chemically complex family of iron-magnesium silicates.

Oxides are another important group of minerals, all of which contain oxygen with nothing else in common. Common oxides include:

1. **Magnetite**—A black iron oxide;
2. **Hematite**—A red iron oxide;
3. **Corundum**—Aluminum oxide with varieties, ruby and sapphire.

Carbonates are a group of minerals, all of which contain carbon and oxygen, with other elements. Important carbonates include:

1. **Calcite**—A calcium carbonate;
2. **Dolomite**—A calcium and magnesium carbonate.

Sulfides all have sulfur in common, with other elements present. Examples include:

1. **Pyrite**—An iron sulfide;
2. **Galena**—A lead sulfide which is a source of lead.

Sulfates all contain sulfur and oxygen with other elements, the most common example of which is **gypsum**, a calcium sulfate.

Native elements are generally quite rare, but very important. Generally, they are pure elements, examples of which include:

1. **Gold.**
2. **Silver.**
3. **Copper.**

Rocks of the Earth's Crust

Rocks are aggregates of minerals. Rocks are subdivided on the basis of where and how they are formed:

1. **Igneous rocks** are those which have cooled from a molten rock mass known as **magma.**
2. **Sedimentary rocks** form at or near the surface by one of two processes:
   a. **Clastic sedimentary rocks** form by the accumulation and solidification of fragments of pre-existing rocks;
   b. **Nonclastic sedimentary rocks** form by the direct precipitation of mineral matter from water (chemical sedimentary rocks) or by the accumulation of plant or animal remains (organic sedimentary rocks).
3. **Metamorphic rocks** are pre-existing rocks which have been changed, usually deep within the crust.

Let's examine each group, in turn.

Igneous rocks have cooled from magma which has been generated deep within the crust or in the upper mantle. Igneous rocks are classified on the basis of:

1. Texture or grain size;
2. Mineralogical composition.

The texture of an igneous rock can usually reveal where the magma has cooled. Magma that is extruded onto the earth's surface cools very quickly, forming very small grains. These are known as **volcanic** igneous rocks. Other magma does not reach the surface; rather, it cools within the crust. Because rocks are poor conductors of heat, magma cooling at depth cools very slowly, forming large crystals. These are known as **plutonic** igneous rocks.

Table 2.3.

Texture \ Mineral Assemblage	Kf>Pl M,A Q	Kf≈Pl A Q	Kf<Pl Pl>A,Px Q?	Px,Ol>Pl	Ol
Volcanic (fine-grained)	rhyolite	dacite	andesite	basalt	
Plutonic (coarse-grained)	granite	granodiorite	diorite	gabbro	peridotite

Kf = potassium feldspar; Pl = plagioclase feldspar; M = mica; A = amphibole; Px = pyroxene; Ol = olivine; Q = quartz

Both volcanic and plutonic igneous rocks are further categorized on the basis of their mineralogical composition. The mineralogical composition is a function of where and under what conditions the magma forms and will be discussed in later chapters. Table 2.3 shows one possible igneous rock classification chart, the one we will use in this book. Note that in all but one case, both volcanic and plutonic counterparts of various igneous rocks are known.

It is important to know that not all volcanic rocks are equally abundant; in fact, the vast majority of volcanic rocks are basalts and andesites. Likewise, not all plutonic rocks are equally abundant; the vast majority are granite and granodiorite. Furthermore, granite is much more abundant than its volcanic counterpart rhyolite, and basalt is much more abundant than its plutonic counterpart gabbro. We believe these relative abundances are a function of magma composition and viscosity. Magma which ends up as either granite or rhyolite is much richer in silica than magma which forms either basalt or gabbros. The amount of silica is apparently what governs the viscosity (flowability) of magma—the more silica in the magma, the more viscous (slower flowing) is the magma and *vice versa*. Thus, silica-poor magma flows relatively easily and tends to make it all the way to the surface to form basalt, rather than pooling at depth to form gabbro. Silica-rich magma, on the other hand, does not flow easily and tends not to make it to the surface to form rhyolite, but pools at depth and forms granite.

Clastic sedimentary rocks form by the accumulation of **sediment,** unconsolidated debris from pre-existing rocks, and **lithification,** the making of sedimentary rock from sediment by compaction and cementation. Clastic sedimentary rocks are sub-classified on the basis of their average grain size:

1. **Shale** is formed by the compaction of clay and mud; shales have an average grain size of less than 1/256 mm.
2. **Siltstones** are comprised of material that has an average grain size between 1/256 mm and 1/16 mm.
3. **Sandstones** are comprised of material that has an average grain size between 1/16 mm and 2 mm.
4. **Conglomerates** are comprised of material that has an average grain size greater than 2 mm and which is generally rounded.

The processes of weathering (breakdown of pre-existing rock), erosion (loosening and removal of sediment from pre-existing rock), transportation (the movement of sediment by wind, water or ice), deposition (removal of sediment from the transporting medium) and lithification (the transformation of unconsolidated sediment into sedimentary rock) are discussed in more detail in Chapter 12 of this book.

Non-clastic sedimentary rocks include **chemical precipitates** from water, such as rock salt, and **organic sedimentary** rocks formed by the accumulation of dead plant or animal debris. Plant material is transformed into **peat** and then **coal;** animals remains generally are comprised of calcium carbonate shell material and are known as **limestones.**

Limestones are formed by three processes:

1. Direct precipitation of calcium carbonate from water, particularly from sea water, in places like Florida Bay and the Bahamas, forming a very fine-grained **oolite;** from fresh water in caves, forming stalagtites and stalagmites
2. Accumulation and cementation of broken and transported shell material, forming **coquina**
3. Colonial growth of organisms into organic reefs, such as **coral reefs.**

Metamorphic rocks are formed by the alteration or transformation of pre-existing rock at depth. The principal agents of metamorphism are:

1. Temperature.
2. Pressure.
3. Chemically active fluids.

To this list, we should add *time*. Metamorphic changes take place while the pre-existing rock is in the solid state, i.e., before melting starts to take place, and most of the chemical reactions and physical changes require a very long period of time.

Metamorphic rocks may undergo *mineralogical changes,* or *textural changes.* Minerals are chemically stable under limited ranges of temperature and pressure, and if those temperatures and/or pressures are exceeded, individual minerals or suites of minerals will change to more stable minerals. Deformational pressures tend to cause minerals to realign themselves in a rock, often transforming a sedimentary or igneous rock that originally had a homogeneous texture (no mineral layering or segregation) into a metamorphic rock with a heterogeneous structure (with mineral layering or segregation).

Any kind of rock can be metamorphosed: igneous, sedimentary or even a previously metamorphosed rock. Table 2.4 gives a list of common igneous and sedimentary rocks and their metamorphic counterparts. Note that some pre-existing rocks, such as shale, undergo a series of transformations into various metamorphic rocks at progressively higher temperatures and pressures. Other pre-existing rocks, such as quartz-rich sandstones, do not undergo such obvious changes.

Table 2.4. Some Unmetamorphosed Rocks and Their Metamorphic Counterparts

Original Rock	Metamorphic Counterpart	Type of Change
Limestone	Marble (all temperatures)	Textural only
Quartz-rich sandstone	Quartzite (all temperatures)	Textural only
Shale	Slate (low temperature)	Mostly textural
	Schist (moderate temperature)	Textural & mineralogical
	Gneiss (high temperature)	Textural & mineralogical
Granite	Gneiss	Mostly textural
Basalt	Greenschist (low temperature)	Textural & mineralogical
	Mica schist (moderate temperature)	Textural & mineralogical
	Amphibolite (high temperature)	Textural & mineralogical

Rock Cycle

All rocks are related to one another and can be transformed into other rocks under the appropriate conditions. The relationship between igneous, sedimentary and metamorphic rocks is known as the **rock cycle,** which is shown diagrammatically in Figure 2.4.

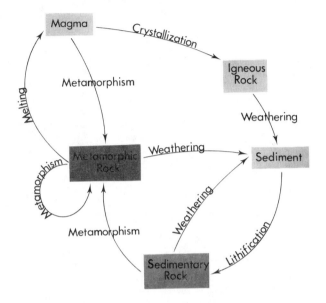

Figure 2.4. The rock cycle.

Note that rock types are shown in boxes; the transformation processes and possible paths are also shown. Note also that almost all transformations are possible, e.g., *any* kind of rock can undergo metamorphism, *any* kind of rock can be weathered to form sedimentary rock. The only prohibited paths are from sedimentary rock to igneous rock (must go through metamorphic rock stage) and igneous rock to igneous rock (must also go through metamorphic rock stage). Be careful *not* to oversimplify the rock cycle. If we start with, say, a basalt and take it through sedimentary and metamorphic changes, until ultimately it is remelted, it will *not* become a basalt again. A whole series of permanent chemical changes take place in the processes that will not allow a basalt to reform.

The conditions under which several kinds of igneous and metamorphic rocks form are discussed in Chapters 9, 10 and 11 of this book, dealing with *internal* processes. The conditions under which several kinds of sedimentary rocks form are discussed in detail in Chapters 12, 13 and 14 of this book, the section dealing with *surface* and *near-surface* processes.

CHAPTER 3
Geologic Time

Geologic time is another basic concept we need to examine. We have alluded to it earlier, but this concept is peculiar to the earth sciences. Many of you are probably astounded, or perhaps numbed, trying to put *billions* of years into a time frame for humans. In this section, we will examine two aspects of geologic time, including:

1. Relative time.
2. Absolute time.

Relative Time

In the field, geologists deal principally with the concept of *relative* time, i.e., the sequence of events which occurred to bring a rock sequence to its present state or position. Relative time does not deal in units of time, e.g., seconds, years; rather it deals with a relative sequence of events.

There exists a series of "principles" that deal with relative time.

1. **Uniformitarianism** is the idea that processes which are at work now have been working throughout geologic time (e.g., gravity-related processes). Uniformitarianism does not extend to the *rates* at which these processes occur. Rates may differ throughout geologic time. This principle allows us to observe geologic processes and products that are happening *now* and extrapolate back into geologic time. For example, consider the beach processes at work today (tides, longshore drift, currents, etc.) and see the products (beaches, barrier islands, dunes etc.). If we see these same, or similar, products preserved in the rock record, we can assume that beach conditions and processes similar to those acting today were very likely active in this place at some previous time.

2. **Superposition** states that when dealing with a layered sequence of sedimentary rocks (or stratified lava flows, for that matter), the *oldest* layer is on the bottom of the stack, and the youngest is on the top, unless the entire sequence has been turned upside down. In many instances, it is not clear "which way is up" in a tilted sequence; in those cases, we can use any of several criteria to determine the tops of sedimentary rocks.

Absolute Time

Absolute time deals in units and implies a method of measuring time. In the past, geologists and others concerned with the establishment of the *exact age* of some event or some thing used a series of dating techniques varying in sophistication. One of the early attempts to measure the age of the earth was undertaken by Bishop Ussher, an Anglican clergyman in the mid-1600's. He used information contained in the Book of Genesis, specifically the lineages of Old Testament families, and calculated that the earth was created on October 26, 4004 B.C. (at 9:00 AM!). This is a far cry from the presently accepted (by geologists) age of the earth of approximately 4.6 billion years; but, in his day, and for some time afterwards, Bishop Ussher's date was accepted.

Other, more scientific, methods for measuring the age of the earth have led to progressively more sophisticated estimates. Many enterprising geologists of the 18th and 19th centuries attempted to determine the age of the earth by calculating how long it took for the oceans to gather the amount of salt contained to date. These calculations were based on estimating the average amount of salt being delivered to the ocean annually and dividing the amount of salt in the sea by the amount of salt delivered each year. They concluded that the earth was approximately 100 million years old. This date was accepted for a considerable length of time.

Lord Kelvin, in the 19th century, attempted a more sophisticated mathematical and physical approach. By that time, it was known that the interior of the earth was giving off heat to the atmosphere, and the rate at which the earth was losing heat had been measured. He made the assumption that the earth was, early in its history, completely molten, and the heat escaping today was residual heat from that originally molten state. He calculated the age of the earth by dividing the total amount of heat originally contained in the molten earth by the rate of heat loss found in his time. His calculations yielded an earth approximately 30 to 40 million years old.

Most of the other 19th century attempts to establish the age of the earth yielded ages of 100 million years or less. So, until very recently, this figure of 100 million years was more or less accepted. Many of these other attempts were based on determinations of how much sediment has accumulated through geologic time and the rate of deposition of sediment today.

Radiometric Dating

In 1896, a French physicist named Becquerel discovered radioactivity, or radioactive decay, the process by which certain chemical elements break down spontaneously to form lighter atomic-weight elements, giving off submicroscopic particles in the process. This discovery provided a new approach known as **radiometric** (measuring radioactivity) **dating** for those geologists who were concerned with establishing the exact, or absolute, age of the earth.

Before looking at *how* this radioactivity is used to determine the absolute age of the earth or, for that matter, the age of many types of rocks, let's look at the process of radioactivity, or radioactive decay. Not all chemical elements are radioactive; most, in fact, are stable and will not spontaneously break down or decay. Generally, the heavy chemical elements are much more likely to decay than lighter elements.

Probably the two most important points to remember concerning radioactive decay are the following:

1. Atoms of the same element always break down in the same way, forming the same decay products. For example, uranium^{238} *always* breaks down to thorium^{234} and a helium nucleus (or *alpha* particle).

2. The *rate of decay* is always the same for a given element.

Let's look at these two points in a bit more detail. To do so, we must know some basic information about the structure of **atoms.** An atom is the fundamental, submicroscopic building block of chemical elements. It was previously thought of as having a structure similar to the structure of our solar system. At the center of the atom, corresponding to the sun, is the **nucleus** of the atom. Surrounding the nucleus, **electrons** were thought to occupy various orbits in space, much like the planets. More modern models of the atom still have a central nucleus, but have it surrounded by electrons in vaguely defined electron clouds. The nucleus is composed of two kinds of subatomic particles, **neutrons** and **protons.** These two particles are approximately the same size and have approximately the same mass. The major difference between them is that the proton carries a **positive electrical charge;** whereas, the neutron is **electrically neutral.** Thus, the nucleus of an atom *always* carries a positive charge, and the amount of charge depends upon the number of protons in the nucleus. The electrons are much smaller and *not* part of the nucleus. They carry a negative electrical charge, and usually there are as many electrons surrounding a nucleus as there are protons in the nucleus.

Next, let's learn a little chemistry shorthand. Earlier, we mentioned that uranium^{238} always breaks down to thorium^{234} and a charged helium nucleus. Let's examine that process in some detail. Uranium^{238} is composed of 92 protons, 146 neutrons and 92 electrons. (*All* forms, or **isotopes,** of uranium have 92 protons and 92 electrons; the number of neutrons can vary.) When uranium^{238} decays, the decay products are thorium^{234} (90 protons and 144 neutrons) and a charged helium nucleus (2 protons and 2 neutrons). Note that *matter,* in the form of neutrons and protons, is neither increased nor decreased; the total number of protons in both decay products is 92, the total number of neutrons is 146. Different isotopes of a given element decay in their own individual way. For now, let's leave this topic and move on to point 2 made earlier, concerning the rate of radioactive decay.

Let's use again our example of uranium^{238}. Radioactive decay may be described as a statistical process in that it is based on probability. Uranium^{238} decays in such a way that it is expected that one-half of the original number of uranium^{238} atoms will spontaneously decay in 4.5 billion years. The amount of time required for one-half of the original atoms of a radioactive element to decay to its decay products is known as its **half-life.** Let's look at this process from the point of view of a hypothetical element x which breaks down to y, shown diagrammatically in Figure 3.1. Let's assume that we start off with 32 atoms of element x, and that the half-life of element x is 10 years. After 10 years, one-half of the original number of atoms of x will have decayed, leaving 16 atoms of x, with 16 atoms of y. After another 10 years, one-half of the remaining 16 atoms of

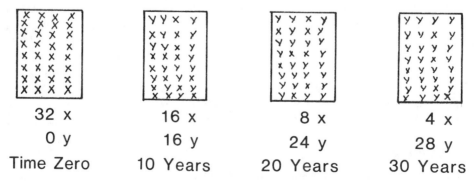

32 x	16 x	8 x	4 x
0 y	16 y	24 y	28 y
Time Zero	10 Years	20 Years	30 Years

Figure 3.1. Concept of half-life. In each container there are 32 atoms, but the number of x atoms is reduced by ½ in each, while the number of y atoms increases in each case.

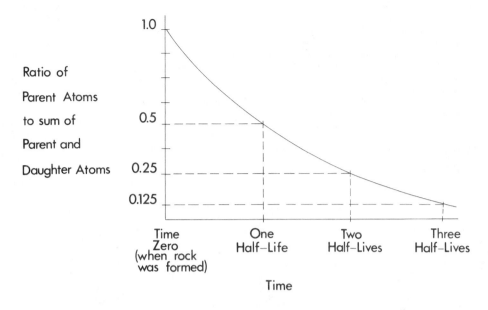

Figure 3.2. Decay curve for parent atom with a half-life of ten years.

x will have decayed, leaving 8 atoms of x and 24 atoms of y. After another 10 years, only 4 atoms of x remain (along with 28 atoms of y). Likewise, every ten years, one-half of the remaining number of atoms of x will be left. (Theoretically, then, we will *never* run out of atoms of x, no matter how long we wait.) This type of decay is shown diagrammatically in Figure 3.2.

We hope by now you can see how we can use that information to determine the age of a rock containing, say x and y, by determining how much of each is in the rock at any given time. This is the basis for **absolute age determinations.** We must make some assumptions, however, before moving on. If we want to determine how old a rock containing x and y is, we have to assume:

1. Only atoms of x were incorporated when the rock was formed or that if some number of atoms of y were incorporated, we must know how many atoms of both were originally incorporated; and
2. No atoms of x or y have been lost from the rock or added to the rock since the time the rock was formed.

Let's examine this process in terms of the products. Assume that a rock formed with 128 atoms of x and no atoms of y. If we pick that rock up after it has been around for ten years, it will have only 64 atoms of x and 64 atoms of y. If we picked the rock up at some unknown time after formation and found that it contained 32 atoms of x and 96 atoms of y, we would know that the rock had been around for 20 years. This decay of radioactive elements can be graphed so that we can determine the age of any rock or mineral containing x and y and which has obeyed the two assumptions.

Now let's see how to use this type of graph. Let's assume that we find a rock containing 6 million atoms of x and *no* atoms of y. The ratio of x to $(x + y)$, $\frac{x}{x + y}$, equals one. We would know that the rock had just been formed. If we found another, similar, rock containing 8 million atoms of x and 16 million atoms of y, we could determine the age of the rock by the following process.

1. Determine the ratio of x to $(x + y)$

$$\frac{8 \text{ million}}{8 \text{ million} + 16 \text{ million}} = \frac{8 \text{ million}}{24 \text{ million}} = 0.33$$

2. Enter on the vertical scale of Figure 3.3 at the ratio, in this case 0.33, and draw a horizontal line to where it intersects the decay curve, then draw a vertical line to the horizontal axis, and this value is the age of the rock.

In this case, the rock is approximately 16 years old.

Remember that this is just a hypothetical example, but there are decay schemes used, which are based on actual chemical elements, actual decay products and actual half-lives. Some of the most widely used are:

1. Uranium-lead (uranium[238] breaks down in a series of decays, ultimately forming lead[206] which is stable), with a half-life of 4.5 billion years.
2. Potassium-argon (potassium[40] breaks down to argon[40]), with a half-life of 1.3 billion years.
3. Rubidium-strontium (rubidium[87] breaks down to strontium[87]), with a half-life of 47 billion years.
4. Carbon[14] (carbon[14] breaks down to nitrogen[14]), with a half-life of 5570 years.

The oldest rock ever studied is estimated to be about 3 or 4 billion years old. Therefore, we have reason to believe that the earth is much older than that. In fact, the age of the earth generally accepted by scientists is

15

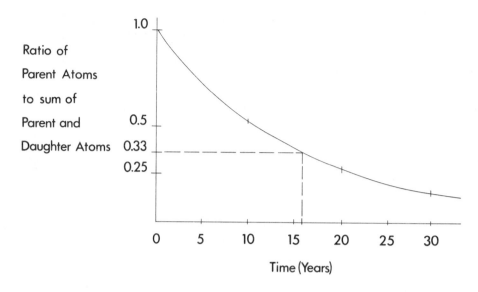

Ratio of
Parent Atoms
to sum of
Parent and
Daughter Atoms

Figure 3.3. Determination of absolute age based on the ratio of parent atoms to sum of parent and daughter atoms. Parent atom has half-life of ten years.

4.6 billion years, with the age of the solar system being considerably older than that, perhaps 10 billion years.

Geologic Time Scale

For many years, even preceding the relatively new era of absolute age dating, geologists have divided and subdivided geologic time into different eras and periods. The original subdivisions were based on fossil animal remains and the relative positions of rock strata containing those animals. Names for the various geologic periods were generally established geographically. A period of time containing a distinctive group or *suite* of fossils was named after the geographic location where rock strata containing these fossils were particularly well exposed. For example, the Permian period was named after the Perm region of Russia, where a sequence of rock strata contained a particularly well-exposed suite of reptile fossils.

The geologic time scale is shown in Figure 3.4. Note that there exist subdivisions of time longer than periods (eras) and subdivisions shorter than periods (epochs). The Paleozoic Era is that time span in which first animal life appeared (although now we know that there exist some rare fossil animal remains in rocks still generally classified as Pre-Cambrian) and in which first vertebrate animals, first land plants, first land animals (amphibians) and first reptiles appeared. The Mesozoic Era is marked by the first mammals, first real birds and first flowering plants. The Mesozoic Era is also that time span in which lived most of the dinosaurs. The Cenozoic Era is marked by the first appearance of manlike primates and extends to the present.

Epochs are subdivisions of periods and are based, in some cases, on geographic locations. In other cases, ep-

ochs are subdivided into Upper, Middle and Lower, e.g., the Upper Triassic Epoch.

The most important subdivision of time was based on what earliest geologists believed to be the presence or absence of fossil animal remains, a boundary based on the presence or absence of life. All of that time (as seen in the rock record) that predated life was designated Pre-Cambrian (before Cambrian).

You will also note that we have listed the absolute time (in terms of millions of years before present) for the beginning of each of these subdivisions of time. Note that the Pre-Cambrian ended approximately 570 million years ago. That means that the Pre-Cambrian comprises approximately 87% of all geologic time. Considerably less is known about the Pre-Cambrian than all of the Cambrian and later periods, because most of our knowledge of geologic history is based on being able to tell something about the conditions under which rocks formed, and those rocks bearing fossils yield considerably more information about these conditions for rock formation (sedimentary rocks) than nonfossiliferous rocks. Until relatively recently, we knew very little about the conditions under which Pre-Cambrian rocks formed. In the past few decades, geologists have begun learning about conditions in the Pre-Cambrian. This information is still relatively hard to come by, and there is considerable difference of opinion among geologists over how to extrapolate backwards in time from conditions that exist now. In other words, the principle of Uniformitarianism, while probably still holding for the Pre-Cambrian, is considerably more suspect when we are extrapolating back a billion or more years than when we are extrapolating back a mere hundred or so million years.

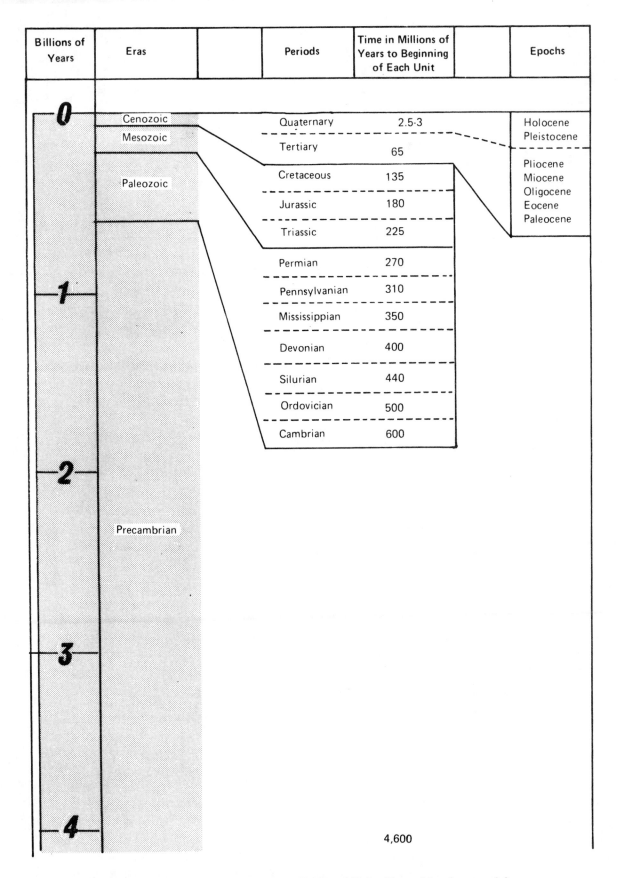

Figure 3.4. The geologic time scale. From Pirkle and Yoho, *Natural Landscapes of the United States*, Third Edition. Kendall/Hunt Publishing Company. Copyright © 1982 by E. C. Pirkle. Reprinted with permission.

CHAPTER 4
History, Development and Early Evidence of the Concepts

The origin and development of the concept of continental drift and its evolution into the concept of plate tectonics provides one of the most remarkable examples of how scientific concepts develop and change. It is replete with brilliant early observations, skeptical nonacceptance, gradual change, rapid change and gradual conversions of early nonbelievers. It is the story of the evolution of a concept that literally revolutionized the earth sciences—as important, perhaps, as the discovery that the earth was not the center of the universe.

Early Observations

In order to understand the history and development of the concept of plate tectonics, it is necessary to go back in time approximately 360 years ago, to 1620. It was at this time that Sir Francis Bacon called attention to the similarities in shape of the eastern coastline of South America and the western coastline of Africa. To the best of our records, this was the earliest notice paid to this physical phenomenon. In 1658, the Frenchman Francois Placet also wrote about this remarkable resemblance. Neither of these early scholars, however, proposed that the continents were two parts of an earlier, larger land mass which had separated. That idea had to wait until 1858. In the interim, the resemblance was considered a curiosity, but, apparently, little was done to develop any cause and effect hypothesis. No doubt, much of this lack of progress stemmed from poor communication and, frankly, a lack of interest on the part of early natural historians. Remember that these observations coincided rather closely with the revolutionary ideas of Copernicus, Galileo and Kepler that the earth was *not* the center of the universe. Consider that the Church at the time condemned these ideas as being heretical and radical and called for the reaffirmation that the earth *was,* in fact, the center of the universe. Is it surprising that an idea so radical as a moving earth "skin" was not considered?

It was not until 1858 that Antonio Snider-Pelligrini suggested that South America and Africa represented pieces of an earlier, larger land mass which had broken unevenly along the present-day coastline. Snider-Pelligrini backed up his contention with paleontological evidence. He identified very similar plant fossils in coal beds, of approximately the same age, in South America and Africa. He contended that this similarity must have been brought about by the juxtaposition of the continental masses, rather than coincidental parallel evolution. His work went basically unheeded, however.

Continental Drift Theory

Not until later, in the 20th century, was the idea of mobile continental masses again brought into the light. In 1912 Alfred Wegener, a German geographer, proposed in his book *The Origins of Continents and Oceans* that *all* of the continents, not just South America and Africa, were once joined in one supercontinent which he called **Pangaea**. Wegener and his colleagues based their proposition on several lines of evidence:

1. Similarity in shapes of coastlines—and an ability to "fit" all of the pieces of the jig-saw puzzle into a coherent picture (Figure 4.1).

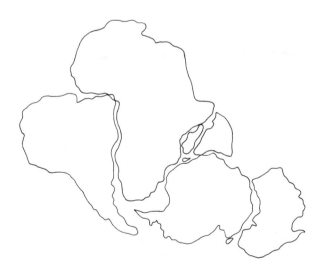

Figure 4.1. The "fit" of southern hemisphere continental land masses into a supercontinent.

2. Fossil similarities—including the fossil plants of Permian age in the present-day Southern Hemisphere and the fossil animal Mesosaurus found in

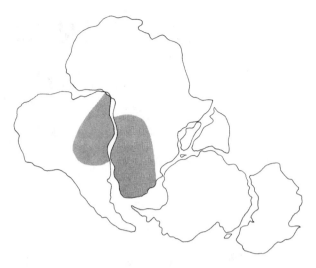

Figure 4.2. Approximate locations of *Mesosaurus* fossils.

both South America and Africa (Fig. 4.2). He used the same kind of logic that had been used by Snider-Pelligrini, that of continental juxtaposition rather than parallel evolution.

3. Pattern of Permo-Carboniferous glaciation—evidence of glaciation approximately 280 million years ago in the present-day Southern Hemisphere and the fact that patterns in glacial movement can be explained by one massive continental ice-sheet moving in consistent directions across a single continental mass (Figure 4.3).

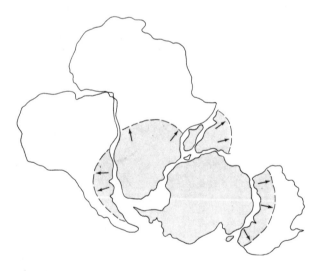

Figure 4.3. Pattern of Permo-Carboniferous glaciation. Arrows indicate direction of movement of ice.

4. Distribution of rock types, ages and structures—the continuity of rock types and rock structures, such as roots of folded mountain belts of the same age, which can be traced from one continental mass to another (Figure 4.4).

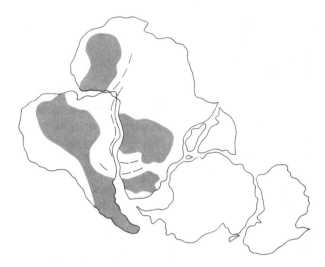

Figure 4.4. Distribution of ancient continental rock masses. Dashed lines indicate general structural trend of folded mountain belts.

All of this evidence pointed to the idea that the continents, as we know them today, were all part of one supercontinent in Permian time, approximately 250 million years ago. Wegener argued that South America and Africa began to split apart approximately 70 million years ago, with the split of North America from Europe occurring only within the past few million years.

Wegener proposed that "drift" occurred as the continents sailed through oceanic crust like rafts driven by the rotation of the earth. This was undoubtedly the weakest link in his chain of arguments for continental drift. Unfortunately, many geologists and especially geophysicists jumped on this weak link, which could be shown to be impossible by examination of information known even then about the strength of rocks.

During the period 1915–1930, a great debate raged concerning continental drift. Those who believed continued to believe and to restate their evidence; those who did not believe continued to disbelieve and to restate their reasons why drift could not occur. We see here a fundamental difference between geologists which, to some extent, continues today. The field of geology is pursued by some people who do field work only, who have little or no use for mathematical/physical theory or manipulation, and by other people who work only on theory (or in a lab) and who do essentially no field work to back up their theoretical models and contentions.

From about 1930 to the early 1950's, the debate on continental drift waned in North America. The "party-line" during that period was that drift could not have taken place, in spite of the field evidence to support the contention. In the Southern Hemisphere, however, and to some extent in Europe, the debate continued, and geologists and geophysicists continued research into the evidence supporting drift.

Physiography of the Ocean Floors

In the 1950's, especially the late 1950's and the 1960's, the scientific community learned more about the physiography of the ocean floor than it had in all the years before. Vast sums of federal money were allocated to mapping the ocean floor. From this research, we acquired much of the information that ultimately led us to that scientific revolution of the mid 1960's—the development of the plate tectonic concept.

There had been a few earlier ocean floor surveys, for example, the surveys by the research vessel *Challenger* in the 1800's, and the United States had done some ocean floor exploration during World War II to support its naval operations in the Atlantic and the Pacific. However, we still had relatively little information about large-scale features. Modern exploration of the ocean floors requires specially rigged ocean-going vessels. At first, old warships were re-rigged for oceanographic research; later, research vessels were specially designed and constructed. In order to produce relief maps, un-

Figure 4.5. Photograph of the Scripps Institute of Oceanography research vessel *Vema*. (Photograph courtesy Willard S. Moore.)

derwater sound waves are generated from shipboard. They bounce or reflect off the bottom and give us a vast amount of information on what the ocean floor looks like. What was learned was that the ocean floor was not flat and featureless, as was previously thought. Rather, it was discovered that the ocean floor was as varied as the land surface above sea level. Two very distinctive features, long volcanic mountain chains and deep, curved or **arcuate trenches,** in particular, were to play a key role in the development of plate tectonic theory in the 1960's. Figure 4.6 is a view of the ocean floor (without the water).

Note that an essentially unbroken volcanic mountain chain begins at Iceland, proceeds south in the Atlantic Ocean, around Africa and north in the Indian Ocean. There, it splits into two arms. One arm apparently converges on the African continent and the other arm continues south and east of Australia and east in the South Pacific until it converges on North America. Note also the remarkable "coincidence" of deep trenches with other volcanic islands, especially around the perimeter of the Pacific Ocean (the Japanese Islands, the Aleutian Islands, the Marianas Islands, and so forth).

When these features were first discovered, there was no all-encompassing theory to explain them. Most geologists saw them as unrelated features. How wrong we were!

Heat Flow Studies

We have known for several hundred years that the earth was losing heat to the atmosphere and hydrosphere. Recall from Chapter 3 that heat-flow studies were underway in the 1800's. In 1846, Lord Kelvin used existing data to try to determine the age of the earth. Throughout the 1950's and 1960's, geologists and geophysicists made more sophisticated determinations of the earth's heat flow over continental and oceanic areas in order to determine the origin of the heat and the magnitude of the heat flow.

Before examining the results of these studies, let's examine the possible sources of the heat and how the heat is transmitted. Recall from Chapter 2 that we believe that the temperature of the earth increases with depth (Figure 2.3). It is now believed that the heat escaping from the earth comes from two sources: (1) "new" heat continually produced by decay of radioactive elements in the crust and mantle and (2) "old" heat from the conversion of gravitational energy to thermal energy in the early stages of earth history (perhaps 4.5–5 billion years ago). It is now mathematically possible to determine, at any point on the earth's surface, the contribution of each to the total heat being lost by the earth.

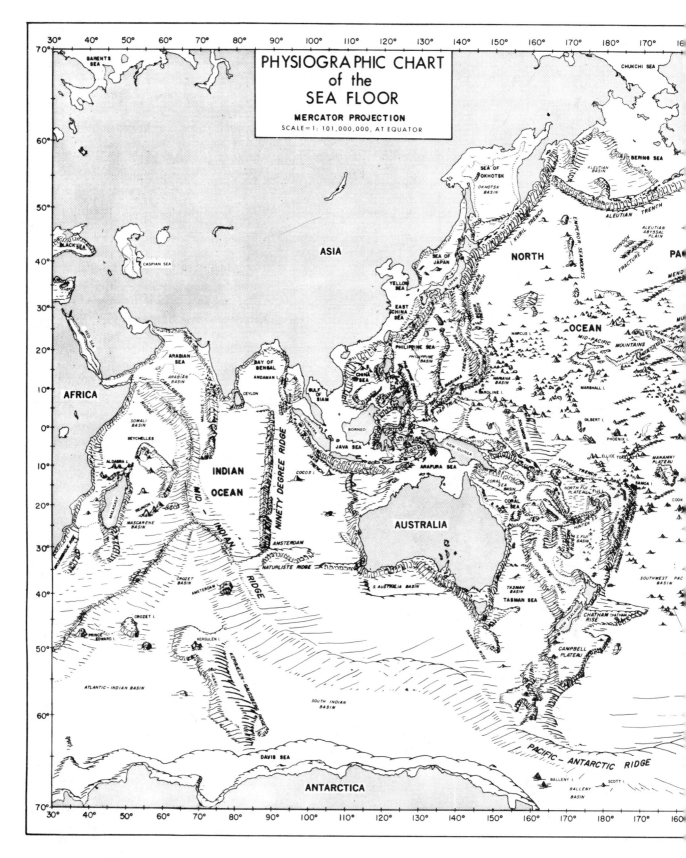

Figure 4.6. Physiographic map of the ocean floor. Courtesy of Hubbard Scientific Co., Northbrook, Illinois.

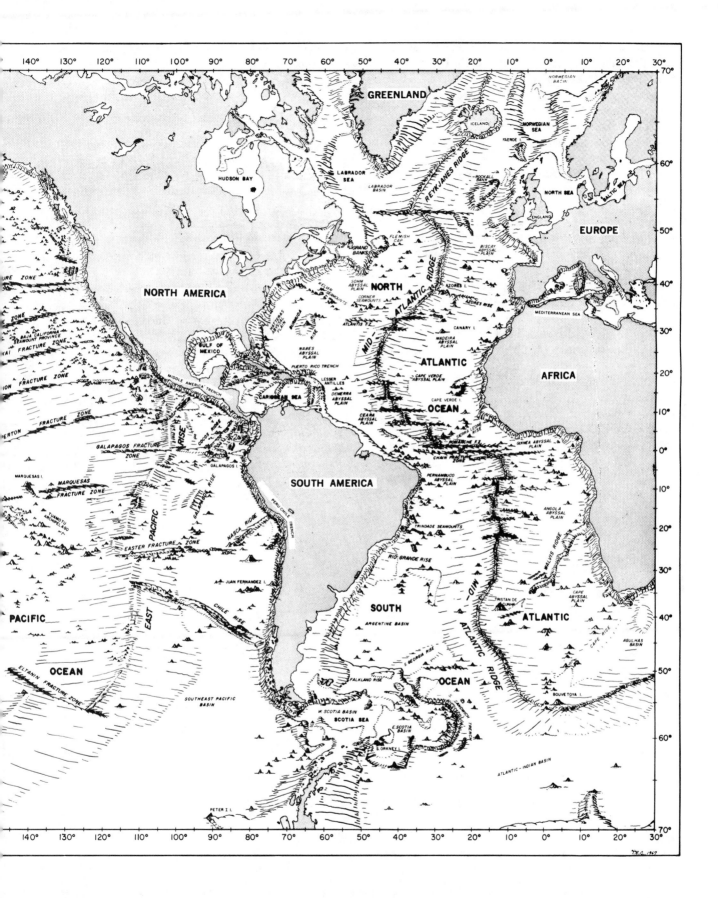

Granite contains more radioactive elements, by far, than any other principal type of rock in the crust or mantle and produces, per gram of rock, as much as six times the heat produced by basalt and about 300 times the heat produced by peridotite. One might guess, then, that higher heat flow values would be obtained over continental areas (essentially granitic in composition) than over oceanic areas (basaltic in composition). In fact, however, average heat flow values over continental areas are about equal to average heat flow values over oceanic areas. Most of the heat being lost by continental rocks apparently comes from radioactive decay. Most of the heat being lost by oceanic rocks must be coming from some other source. It is now thought that the source of the heat being lost by oceanic rocks is the heat that is being given up by the cooling of the continually forming basalt that constitutes the crust in the oceanic areas.

Even though the *average* heat flow values over the earth are about equal, some systematic variations in heat flow were found that had to be explained. It was found, for example, that heat flow values were higher over the ocean ridges (not surprising since these have long been known to be volcanic in origin) and lower over the trench areas (for reasons that were not to become clear until the development of plate tectonic theory) and about "average" over the other parts of the ocean floor. These observations mystified oceanogra-

phers in the 1950's and early 1960's. With the advent of plate tectonics theory, we will see that these variations are not only explainable, but predictable.

Gravity Studies

Studies of the earth's gravitational attraction also began in earnest in the 1950's and 1960's, and these studies revealed some additional curiosities about the earth's crust and mantle that could not be explained by then-current ideas on the permanency of crustal land masses. Before getting into these studies and these curiosities, it is necessary to introduce another concept, the concept of **isostacy,** in order to understand the implications of the gravity studies.

According to the concept of isostasy, the crust of the earth, both continental and oceanic, behaves as if it were *floating* on the underlying mantle. Further, in this concept, there is a depth within the mantle known as the **isostatic compensation depth** where the weight of all of the overlying rock material is equal (see Figure 4.7). If the earth were in isostatic equilibrium, where all parts of the crust were floating at the level they should, then the isostatic compensation level would be at a uniform depth within the mantle. This could only occur if the lithosphere were static (not moving). As you already know, the lithosphere is *not* static. Rather, plates are

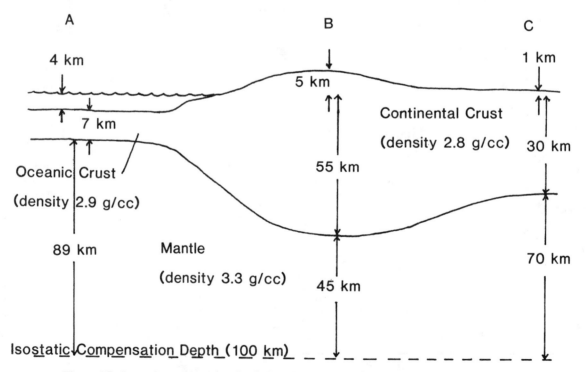

Figure 4.7. Isostatic compensation depth. Based on crustal thicknesses and densities given in this diagram, the mass of a 1-cm diameter column of rock at points A, B, C will be equal at a depth of 100 km. Reprinted with permission of Macmillan Publishing Co., Inc., from *The Evolving Earth,* Second Edition, by F. J. Sawkins, C. G. Chase, D. G. Darby, and G. Rapp, Jr. Copyright © 1978 by Macmillan Publishing Co., Inc.

in constant motion, and we will see later that this motion causes the isostatic compensation depth to vary.

Isostatic equilibrium would cause the continental crust (with a relatively low density) to be thicker than oceanic crust (with a higher density). This variation in thickness was verified by seismic studies showing that, in general, continental crust is thicker than oceanic crust. Further, the higher continental crust stands above sea level, the deeper the crust extends into the mantle.

Gravity studies over the surface of the earth are based on the gravitational attraction between a mass and the earth at various locations on the earth's surface. The gravitational attraction between any two objects is given by the formula:

$$F = \frac{G\,(m_1)\,(m_2)}{r^2},$$

where F is the gravitational attraction, G is the universal gravitational constant, having a value of 6.67×10^{-8}, m_1 and m_2 are the masses of the two objects, and r is the distance between the centers of gravity of the two objects. Gravitational measurements involving the earth are made on the surface of the earth by weighing an object at various locations. The **weight** of an object varies from place to place on the earth's surface. It is equal to the object's **mass** (a constant) times its acceleration due to gravity (**g**), a value which is variable but with a value of about 980 cm/sec². In this case, m_1 becomes the mass of the object being weighed, and m_2 becomes the mass of the earth (M_e). The weight of the object (F) is given by:

$$F = (m_1)(g).$$

The earlier formula then becomes:

$$F = m\,g = G\,\frac{m_1 M_e}{R^2}$$

(where R is the radius of the earth). This formula can be simplified to:

$$g = G\,\frac{M_e}{R^2}.$$

An object then is weighed at several locations. The *weight* is then divided by the *mass* to determine *g*. A nonrotating spherical earth should then show equal gravitational values (values of g) at any point on its surface. However, the earth is not spherical but *elliptical,* with a greater radius at the equator than at the poles. As a result, one might expect gravitational readings to be lower at the equator than at the poles (remember the gravitational attraction is *inversely* proportional to the square of the distance between the two objects.) Further, because the earth is rotating, there is a centrifugal force that is tending to throw objects *away* from the earth. This centrifugal force, which is greatest at the equator and zero at the poles, reduces the gravitational attraction. The "bottom line" is that things weigh *less* at the equator than at the poles.

Surface topography also affects gravity readings. Readings on a high mountain are made farther from the earth's center than readings at sea level.

All of these complications, however, can be corrected for, and, when the corrections were made, geophysicists *still saw* variations in the gravity readings. These variations are due to differences in the density of rock in the earth's crust and are known as **gravity anomalies. Negative gravity anomalies** occur over less dense rock, and positive gravity anomalies occur over more dense rock. You should be able to deduce that, in general, negative anomalies were seen over continental areas, and positive anomalies were seen over oceanic areas.

If these were the only gravity anomalies which had been found, they would still be important in terms of understanding compositional differences in the earth's crust, but they would not have any bearing on plate tectonics. However, this is not the end of the story. Marine geophysicists, those specializing in the geophysics of oceanic areas, found a curious gravity feature not explainable by compositional variations alone. They found that all deep oceanic trenches were characterized by large *negative* gravity anomalies. (Recall that most oceanic areas were characterized by positive anomalies.) What caused these negative anomalies? Not compositional differences—the oceanic crust in the trench areas was basaltic just as in other trench areas. The suggestion was made in the late 1950's (prior to the plate tectonic theory) that these were areas where oceanic crustal rocks were either being continually pushed down or pulled down into the mantle. However, no mechanism for crustal movement was universally accepted at that time, and the idea did not really "catch on." But negative gravity anomalies over the oceanic trenches did point up once again, as had low heat flow from the trenches, that those trenches were unusual areas.

CHAPTER 5
Earthquakes and Seismicity

The study of earthquakes has been going on for literally hundreds of years. No wonder! Earthquakes are the most destructive forms of energy release known to mankind. While earthquakes can, and do, occur almost anywhere, the vast majority of earthquakes take place in geographically restricted zones. This has been known for many years prior to the development of the plate tectonics model, but only since the advent of this new model do we understand *why* earthquakes occur where they do. The earthquake zones so clearly delineated in Figure 5.1 are now seen to be plate margins.

But first, what exactly is an earthquake? By definition, an earthquake is movement of the ground due to energy being released by the movement of large rock masses past one another along a particular kind of fracture known as a fault. Earthquakes come in all sizes, from minor tremors not even felt directly over the place where the rupture took place, to major tremors that are felt for thousands of miles and cause tremendous destruction and loss of life. The underground spot, or zone, where the rupture takes place is known as the **focus.**

The **epicenter** of an earthquake is the spot on the earth's surface that lies directly above the focus. The energy that is released in an earthquake travels in the form of **seismic (sound) waves.**

Seismology, the study of earthquakes and the seismic waves generated by earthquakes, has permitted us to piece together a picture of the internal structure of the earth. The study of seismic waves can tell us the depth at which an earthquake takes place, the amount of energy released by the earthquake and the direction that the two ruptured rock masses moved along the fault. With this type of information, we have learned a great deal about how the internal stresses are directed within the earth and how rocks respond to these stresses. Seismic studies have provided information that has resulted in a clearer picture of the movement of huge rock masses past one another at what we now understand to be plate margins.

Earthquakes vary in intensity, depending on the amount of energy released by the rock rupture. There are two scales on which earthquakes are measured. The

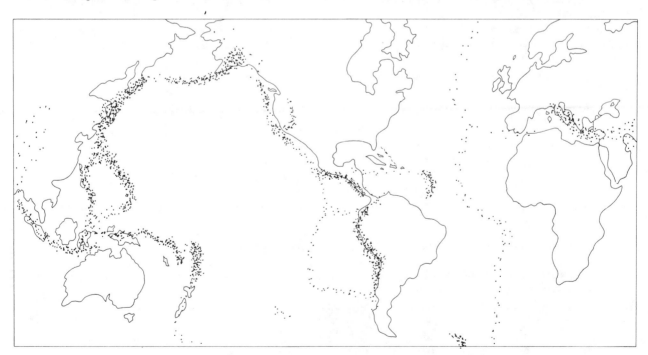

Figure 5.1. World map showing areas of recurring and frequent earthquakes. Compare this distribution of earthquakes to the plate boundaries of Figure 1.1. After James R. Heirtzler, "Sea-Floor Spreading," *Scientific American,* December 1968, pp. 64–65. Copyright © 1976 by W. H. Freeman and Company, Publishers, San Francisco. Used with permission.

Richter Scale is based on the actual amount of energy released and is graduated from a reading of less than 1 to greater than 8. The scale is logarithmic; each unit increase represents a multiplication of 30 times the energy released by the step below it. Examples of well-known earthquakes, with their Richter readings, are given in Table 5.1.

Table 5.1. Some Great Earthquakes

Location and date	Richter Magnitude
Alaska, 1899	8.6
Alaska, 1964	8.6
Colombia, 1906	8.6
Pakistan, 1950	8.6
China, 1920	8.5
Japan, 1933	8.5
Chile, 1906	8.4
Chile, 1960	8.4
U.S.S.R., 1911	8.4
San Francisco, 1906	8.3
China, 1976	8.2
Peru, 1970	7.8
Charleston (SC), 1886	6.8

Another measure of the strength of an earthquake is given by the **Modified Mercalli Scale,** which is based on how surface conditions respond to an earthquake. A description of the Modified Mercalli Scale is given in Table 5.2. While easier to use, the Modified Mercalli Scale is really not as quantitative as the Richter Scale and can be misleading. Two earthquakes giving off the same amount of energy and, therefore, having the same Richter reading can be assigned very different Mercalli readings if the two earthquakes had their foci at two different depths.

Seismograph Recordings

Let's now examine the basic tools of the seismologist. As discussed earlier, earthquakes send seismic waves outward in all directions in all directions from the earthquake focus. These waves or motions can be recorded by a **seismograph,** a device used by geophysicists to study earthquakes. Seismographs are very sensitive and can detect movement unable to be felt by people. A seismograph works on very simple principles, even though modern seismographs are very complex. An early seismograph consisted of a rigid frame securely coupled to the earth. On this frame, smoked

Table 5.2. Modified Mercalli Scale

I Not felt except by a very few under especially favorable circumstances.

II Felt only by a few persons at rest, especially on upper floors of buildings. Delicately suspended objects may swing.

III Felt quite noticeably indoors, especially on upper floors of buildings, but many people do not recognize it as an earthquake. Standing motor cars may rock slightly. Vibration like passing of truck. Duration estimated.

IV During the day felt indoors by many, outdoors by few. At night some awakened. Dishes, windows, doors disturbed; walls make cracking sound. Sensation like heavy truck striking building. Standing motor cars rocked noticeably.

V Felt by nearly everyone, many awakened. Some dishes, windows, etc., broken; a few instances of cracked plaster; unstable objects overturned. Disturbances of trees, poles, and other tall objects sometimes noticed. Pendulum clocks may stop.

VI Everybody runs outdoors. Damage negligible in buildings of good design and construction; slight to moderate in well-built ordinary structures; considerable in poorly built or badly designed structures; some chimneys broken. Noticed by persons driving motor cars.

VIII Damage slight in specially designed structures; considerable in ordinary substantial buildings, with partial collapse; great in poorly built structures. Panel walls thrown out of frame structures. Fall of chimneys, factory stacks, columns, monuments, walls. Heavy furniture overturned. Sand and mud ejected in small amounts. Changes in well water. Persons driving motor cars disturbed.

IX Damage considerable in specially designed structures; well-designed frame structures thrown out of plumb; great in substantial buildings, with partial collapse. Buildings shifted off foundations. Ground cracked conspicuously. Underground pipes broken.

X Some well-built wooden structures destroyed; most masonry and frame structures destroyed with foundations; ground badly cracked. Rails bent. Landslides considerable from river banks and steep slopes. Shifted sand and mud. Water splashed (slopped) over banks.

XI Few, if any, (masonry) structures remain standing. Bridges destroyed. Broad fissures in ground. Underground pipelines completely out of service. Earth slumps and land slips in soft ground. Rails bent greatly.

XII Damage total. Practically all works of construction are damaged greatly or destroyed. Waves seen on ground surface. Lines of sight and level are distorted. Objects are thrown upward into the air.

paper was rolled at a very precise rate. Above this frame, but not attached to it, was a suspended weight to which a marker was attached. When an earthquake occurred, the seismic waves caused the rigid frame to move. The suspended weight tended *not* to move due to its inertia. As the frame and rolled smoked paper moved, the marker attached to the suspended weight caused a line to be recorded on the smoked paper. The line was not straight, but rather was a "squiggly" line marking the movement of the frame due to the movement of the seismic waves. Some seismographs still used a smoked paper, but most use marking pens on plain paper. Figure 5.2 shows a **seismogram,** the trace of the

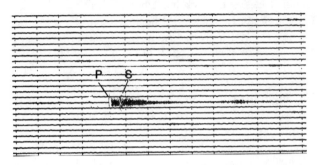

Figure 5.2. Seismogram for a minor earthquake showing first arrival of P and S waves. (Courtesy Pradeep Talwani.)

squiggly line on smoked paper, for a typical minor earthquake. Note that there is no straight line anywhere on the seismogram. Intervals of time are precisely marked on the smoked paper, so that we can determine the exact time at which the seismic wave arrived at the seismographic station. Seismographs are stationed all over the surface of the earth.

There are three types of seismic waves. **P waves** (primary) and **S waves** (secondary) are body waves; they travel within the earth. The third type of wave is the surface wave. While surface waves are as important as P and S waves to the geophysicist, their usefulness is beyond the scope of this book. We will concentrate on P and S waves and what information they can provide us.

P waves are longitudinal or push waves; they cause pulses of higher density (**compression**) and lower density (**rarefactions**) in the material through which they move. These pulses move in the direction the wave is moving. S waves are transverse or shear; in these waves, vibrations are at right angles to the direction of wave movement. Figure 5.3 shows graphically the difference between P and S waves.

Another way to "see" the difference between P and S waves is by way of an analogy. Most of you at one time or another have played with a "Slinky," a coiled metal spring that can "climb" down stairs one at a time. Visualize a Slinky slightly stretched horizontally between two boards. If you were to hit one of the boards sharply, say with a hammer, the force will be trans-

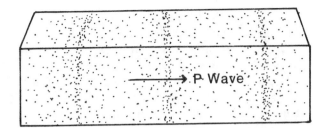

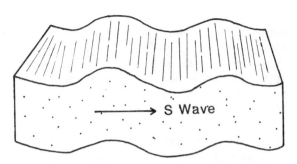

Figure 5.3. Seismic waves. Dot pattern in illustration shows the density of matter. Note variations in density with P waves and consistency of density with S waves. Direction of wave propagation is the same for both types of waves.

mitted to the Slinky, which would respond by compressing a part of itself. This compressed part will appear to move along the Slinky to the other board. In fact, though, no part of the Slinky moves more than a fraction of an inch. Rather, it is energy that is being transmitted, and this energy transmission causes a "wave" of compression to move along the Slinky. This is analogous to a P wave. The Slinky vibrates in the direction the wave moves (horizontally).

We can also use the Slinky to demonstrate an S wave. With the same set-up (horizontal Slinky slightly stretched between two boards), hit *down* on the Slinky. This will cause a wave to start, much like popping a rope or a whip. In this case, the Slinky moves *up* and *down* along its length, with a wave moving in the direction the Slinky is stretched (horizontal). Again, no part of the Slinky is moving very far, but energy can be transmitted for great distances. This is analogous to an S wave. The wave motion is perpendicular to the direction the wave moves.

Both P waves and S waves travel through solids at velocities that are dependent primarily on the density of the material. Therefore, both P and S waves travel at different velocities in different kinds of rocks. Only P waves can be transmitted through liquids. The non-transmission of S waves within the earth's core led geophysicists to suggest that the core was liquid. More recently, careful examination of deep earthquake wave transmission has led them to suggest that only the *outer* part of the core is liquid, and that the *inner* part of the core is *solid*.

The time required for an earthquake wave to reach a seismograph station tells the seismologist, in general, the type of rocks through which it travelled. It is possible to tell, for example, that a seismic wave has travelled through two or more vastly different kinds of rock material on its way to the seismograph station. Furthermore, it is possible to determine where these two (or more) rocks meet. Such a boundary is known as a **seismic velocity discontinuity** (to be discussed in Chapter 7).

Both P and S waves are recorded on seismograms such as the one shown in Figure 5.2. P waves travel faster than S wave, however, and reach the seismograph station earlier. The distance between the first P wave on a seismogram and the first S wave on a seismogram tells us how far the waves have travelled from the earthquake focus. Therefore, this information can give us the distance from the earthquake to a given station.

While P-wave–S-wave differences can tell us distance from an earthquake, they cannot tell us direction. Three nonaligned seismographs can, however, pinpoint an earthquake epicenter. Let's look at an "ideal" earthquake and its effect on three seismograph stations. The stations are shown on the map in Figure 5.4. Let us assume that P-wave–S-wave differences at

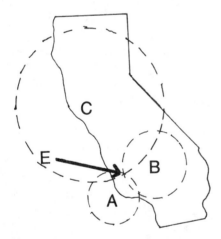

Figure 5.4. Determining the epicenter of an earthquake. The three stations A, B and C are at different distances from the epicenter. Intersection, E, of the three circles marks the epicenter.

the three stations tell us that the earthquake epicenter is 100 miles from seismograph station A, 125 miles from seismograph station B and 175 miles from seismograph station C. We can draw circles around each of these stations, corresponding to these distances. These three circles will intersect at a point, and this point locates very accurately the earthquake epicenter.

With these tools, seismologists can tell the location of an earthquake's focus. Recall that an earthquake is due to the sudden movement of rock masses past one another along a fault. The delineation of earthquake zones, then, is also the delineation of zones along which rocks are moving past one another. We now see that these zones, in fact, *define* plate boundaries. Further, it is possible to determine the relative motion of the plates past one another. In other words, we can tell if the movement is predominantly vertical, predominantly horizontal, or both vertical and horizontal. What we have discovered is that some earthquake zones, like the Mid-Atlantic Ridge and Eastern Pacific Rise, are characterized by rock movements (faulting) that are predominantly vertical. Other earthquake zones, like along the San Andreas fault in California, are characterized by faulting that is predominantly horizontal. Still other earthquake zones, like those associated with the Peru-Chile Trench and Andes, are characterized by faulting that is inclined to the earth's surface, where movement is partly vertical and partly horizontal. This information led us to believe that each of these earthquake zones (plate boundaries) was fundamentally different, and each characterized a particular kind of plate boundary. This was a major breakthrough in the evolution of the concept of plate tectonics. In Chapters 9 and 10, we will discuss these plate boundaries in much more detail and discuss the type of plate motion that is involved with each of these types of boundaries.

Intraplate Earthquakes

The mapping of literally thousands of earthquakes has given us the distribution picture seen earlier in Figure 5.1. From this picture, it is very clear that most of the earthquakes that occur do so along plate margins. However, not all earthquakes are located along plate margins. Some occur within the plates and are known as **intraplate** earthquakes. These are of some interest because two of the major earthquakes that occurred in the United States in the last 200 years have been the Charleston (South Carolina) earthquake of 1886 and the New Madrid (Illinois-Missouri) earthquakes of 1812–1814. Intraplate earthquakes apparently are not related to plate tectonics. Two of these earthquakes were greater than Richter magnitude 8, and both caused extensive property damage and resulted in the loss of many lives. Both of these earthquakes were due to movement along existing faults, but neither of these faults were related to plate margins. While quantitatively not as important as plate margin earthquakes, intraplate earthquakes still can be major. Anywhere there exists stored-up energy in the rock layers, there is the possibility of earthquakes.

CHAPTER 6

Paleomagnetism and Oceanic Magnetic Strips— the Clinching Evidence

Paleomagnetism means, literally, ancient magnetism and involves the study of the magnetic properties of ancient rocks. Before getting too deeply involved in paleomagnetism, it is important to study the physical phenomenon of magnetism, the earth's magnetic field, and how rocks respond to the earth's magnetic field. After that, we can investigate evidence, *locked in the rocks,* of changes in the earth's magnetic field—changes in intensity of the field, polarity of the field, and the location of the magnetic poles of the earth.

Magnetism

Virtually everyone, at some time, has played with a suspended bar magnet and noted that it always come to rest pointing in a north-south direction. Indeed, a compass needle is a thin suspended bar magnet. But what is magnetism, and what is its cause? Magnetism is a property of matter that resides in the atomic structure: How the atoms in a particular crystalline solid are arranged relative to one another and how the electron fields of the various atoms are aligned determine the magnetic strength. Not all materials are magnetic; in fact, relatively few are. Most materials that are magnetic contain iron and, conversely, most crystalline solids that contain iron either are magnetic, or they can be magnetized.

Any magnetic body has associated with it a **magnetic field.** A field is a difficult phenomenon to describe because you cannot see it. Perhaps you recall sprinkling iron filings on a piece of paper and holding a magnet under the paper. The alignment of the iron filings followed the magnetic field. Another good way to come to some understanding of a magnetic field is to remember that a magnet has the ability to attract an object without actually coming in contact with it. It is the magnetic field of a magnet which causes another object to become temporarily magnetized. Only when this second object becomes magnetized can it be physically attracted to the magnet.

A permanent magnet is an object that has its atoms and electron fields "frozen" into a particular pattern.

Other nonmagnetic, but magnetizable, materials do *not* have their atoms and electron fields frozen into that particular pattern, but their atoms and electron fields are alterable so that they can take on, temporarily, the necessary pattern when placed in a magnetic field.

Earth's Magnetic Field

The earth behaves as though it were a simple bar magnet oriented north and south through the center of the earth. However, although the earth's magnetic field causes a compass needle to point north, we certainly will not find a gigantic pole-to-pole bar magnet passing through the earth's center. The exact cause for the earth's magnetic field is *unknown;* but we understand its effect, and we can predict, to a large extent, its behavior. Thus, we can work with it and use it to our benefit, even though we don't know what causes it.

Let's examine the earth's magnetic field a bit more closely. First, we must make a distinction between the earth's rotational pole and its magnetic pole. In general, the magnetic pole is geographically very close to the rotational pole, as it is today, but they may never have corresponded *exactly.* We know that the earth's field is such that the strength of the field decreases away from the magnetic poles of the earth and is at a minimum at the equator. Figure 6.1 shows the earth's magnetic field. If we look at this field closely, it should be apparent that if a compass needle were free to move in a vertical as well as in a horizontal plane, the needle would not only point north, but it would also point down toward the pole of the hypothetical bar magnet along the lines of magnetic force. The direction a magnet points in the horizontal plane is known as the **declination.** The downward dip is known as the **inclination.** At the equator, the needle would point to the north (declination) and would be horizontal (zero inclination). At the magnetic north pole, the needle would point straight down (90° inclination). At various places between the equator and north pole, the needle would point north, and the inclination would depend upon how far from the equator the compass is located. In fact,

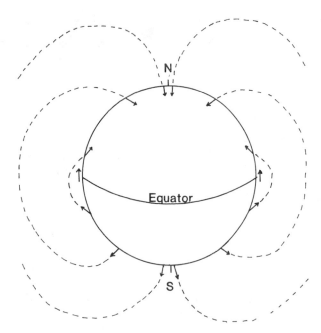

Figure 6.1. The earth's magnetic field. Lines are lines of force, showing the direction of the magnetic field and the inclination a compass needle would take at various points on the earth's surface.

and this is very important to understand, the angle at which the needle would dip turns out to be directly related to the latitude at which it is located. Thus, if we are at latitude 30°N (30 degree north of the equator), the compass needle would dip at an angle that could be converted to 30° by a simple mathematical formula. The importance of this phenomenon will become apparent in the subsequent sections of this book.

Rock Magnetism

Certain minerals, generally iron-bearing, are capable of becoming magnetized. This is possible because a mineral, by definition, is a solid in which component atoms are arranged in a set pattern relative to one another. The electrons associated with iron atoms are generally not permanently aligned in a state of magnetism; nevertheless, in the presence of a magnetic field, they are capable of temporary rearrangement in a pattern that imparts magnetism to the mineral. In general, the strength of the mineral's magnetic field is related to the amount of iron in the mineral. Some minerals, for instance magnetite (iron oxide, Fe_3O_4), are permanently magnetized and have associated magnetic fields.

Let's now consider the conditions under which permanent magnetism can occur. Some rocks originate as magma, a molten, chemically complex mixture of many different kinds of atoms, always including oxygen, silicon, aluminum, calcium, magnesium and iron, as well as others. (Most of you are familiar with two kinds of rock that originate as magma: granite and basalt.) As

magma cools, certain atoms bind together to form minerals in a predictable sequence. Because different substances have different melting points (the temperature at which a solid changes to a liquid state, or vice versa), various minerals in a magma do not solidify at a single temperature. Rather, solidification takes place over a range of temperatures. There is frequently more than 100°C between the highest and lowest melting points. As a generalization, the iron-bearing minerals form early in the cooling process, at rather high temperatures. The iron atoms are fixed in these minerals in characteristic relationships to other atoms. However, their electron fields are not yet arranged in a magnetic pattern, because iron atoms do not respond to the earth's magnetic field at high temperatures.

As the mineral cools, it reaches a temperature known as the **Curie temperature,** a temperature that varies for different minerals, but is usually around 500°–600°C. This is the temperature at which the mineral's electron fields become frozen so as to impart a weak permanent magnetism known as the **remanent magnetism** to the mineral. The electron fields are such that the north-seeking magnetic pole of the mineral aligns itself toward the earth's north magnetic pole. Because these electron fields in magnetic minerals generally align themselves in three dimensions, the mineral becomes aligned so as to point north and dip at the appropriate angle according to the earth's magnetic field. Thus, a study of the declination and inclination of magnetic minerals can tell us the relative position of the north magnetic pole at the time the minerals reached their Curie point. This, we will see, is of extraordinary importance to geologists trying to determine relative locations and possible movements of land masses at specific times in the past. But we are getting ahead of our story. Let's look at this rock magnetism in a historical context.

In the 1950's, researchers in England and the United States began studying magnetism in ancient rocks. A lot of the very important early work was done in the laboratory of the British professors S. K. Runcorn and P. M. S. Blackett at a time when very little work on rock magnetism was being done elsewhere. (Some pioneering workers in Japan and France were pursuing information about paleomagnetism, but not with the fervor of the British scientists.)

Runcorn and his associates and Blackett and his associates examined several layered volcanic lava flows (and some associated layered sedimentary rocks). Using the concepts of declination and inclination explained earlier, they suggested that the north magnetic pole has not always been where it is today. (Remember that "plate tectonics" had not yet been postulated, and "continental drift" was in general disrepute.) The groups determined the declination and inclination based on the remanent magnetism. Using two rocks of the

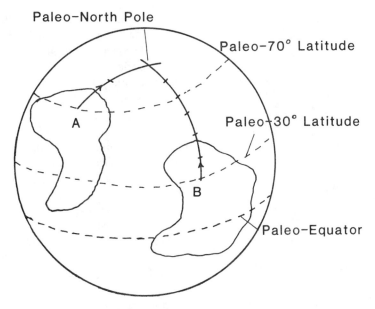

Figure 6.2. Determining ancient magnetic pole positions. Rocks of same age collected at points A and B show different declinations and inclinations. Intersection of the two declinations determines the ancient (paleo) North Pole. Inclinations indicate ancient (paleo) latitude lines of the sample locations.

same absolute age, geographically separated by some distance, they could determine where the north magnetic pole was, relative to where the rocks were collected, at the time the rocks were formed.

Let's look at a hypothetical example of how this might be done. Figure 6.2 shows where two lava rocks with the same ages of "x" years are collected. Prior to moving them, the exact position of the rocks and the relationship between the rocks and the present north magnetic pole must be marked on the samples. In the laboratory, the remanent magnetism of these two samples is determined. The declination of the remanent magnetism indicates the direction from both rocks to the paleo north pole (the position of the pole at the time the rocks were formed). The locations at which the rock samples were taken are noted on a map. The declination of their remanent magnetism is marked. The paleo pole is then found by drawing lines on the map from the sample sites in the direction of the declinations. Where these two lines cross is the location of the paleo north pole. The inclinations of the remanent magnetism tell at what latitude each sample was located with respect to the paleo pole when it formed. We can now locate the paleo equator at time "x" and any other latitude by the following scheme: If the inclination data of the rock from locality B indicates it was on a latitude of 30 degrees at time "x," and if the inclination data of the rock from locality A indicates it was on a latitude of 70 degrees at time "x," we simply draw a line between B and the paleo pole. We divide this line into six equal 10° segments, from 30° at point B to 90° at the paleo pole and find where 70° is located on this line. We now have two points on our map that corre-

spond to the 70° latitude at time "x," point A and the point on the line between B and the paleo pole we have just determined. A line drawn between these two points corresponds to the 70° latitude. Other lines can then be drawn parallel to the 70° latitude line to mark other latitudes at time "x."

Runcorn and his colleagues made several such determinations of past polar positions based on rocks of various geologic ages from England and Europe. What they found was that the magnetic poles apparently have not always been where they are today, at least with respect to England and Europe. Rather, the poles appeared to have migrated or wandered. For example, Runcorn found that during Cambrian time, approximately 500 million years ago, the north pole was apparently located in what is now the Pacific Ocean. With time, it "migrated" northwesterly until, in Permian time approximately 250 million years ago, the north pole was located just north of the Japanese islands. From Permian time to present, the north pole apparently "wandered" toward its present position in the Arctic Ocean.

Runcorn and his associates also studied the paleomagnetism of similar rocks from North America. They found that, according to evidence from North American rocks, some of their ideas about polar wandering held. The north magnetic pole seemed to wander from Cambrian time to present. However, the "path" deduced from data on English and European rocks was similar to, but not identical with, the path deduced from data on North American rocks. Figure 6.3 compares the two apparent polar wandering paths over the past 600 million years.

33

Figure 6.3. Apparent wandering paths of ancient North Pole positions through time, for samples from North America and England-Europe. Reprinted with permission of Macmillan Publishing Co., Inc., from *The Evolving Earth*, Second Edition, by F. J. Sawkins, C. G. Chase, D. G. Darby, and G. Rapp, Jr. Copyright © 1978 by Macmillan Publishing Co., Inc.

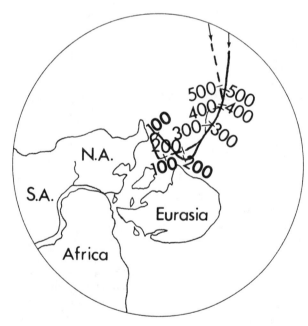

Figure 6.4. Coincidence of North American and England-European polar wandering paths for the period 450 million years BP to 200 million years BP, the period of time the two continental masses were apparently welded together. Divergence of the paths prior to 450 million years BP and since 200 million years BP indicate the continental masses were separated at those times. Reprinted with permission of Macmillan Publishing Co., Inc., from *The Evolving Earth*, Second Edition, by F. J. Sawkins, C. G. Chase, D. G. Darby, and G. Rappy, Jr. Copyright © 1978 by Macmillan Publishing Co., Inc.

Using these data which at first appear contradictory, Runcorn revived the old continental drift controversy. He argued that the two apparent polar paths would overlap almost perfectly, if the eastern North American and western European coastlines were in contact with each other (see Figure 6.4). This seemed to hold true only for the period of time beginning in the late Paleozoic and ending in the early Mesozoic (beginning 450 million years before present and ending 200 million years before present). The polar paths apparently do *not* overlap after the Triassic period (200 million years before present time up to present time). It was found, however, that the paths would overlap if it was postulated that the continental masses began to separate after the Triassic and move toward their present locations.

Evidently, there has been *relative* movement between the north magnetic pole and England-Europe on the one hand, and North America on the other hand. The question then can be asked: Which moved *absolutely*, the magnetic pole as Runcorn had suggested earlier, or the continental masses, or both? The deduction that England-Europe had moved relative to North America and that North America had also moved relative to the north magnetic pole strongly suggests that the continental masses have moved (and possibly continue to move absolutely). The question about absolute movement of the north pole is still not totally resolved. As we will discuss next, there is evidence that the north *magnetic* pole is moving relative to the north *rotational* pole, but only to a small degree.

Polarity Reversals

We now come to a magnetic phenomenon that has proved to be very important to the story of plate tectonics. It seems that the north-south polarity of the earth's magnetic poles may have reversed periodically. No one yet knows how or why this happened, but there is essentially no quarrel among geologists or geophysicists that this phenomenon has, in fact, taken place and may occur again in the future.

Before getting into the relationship between polarity reversal and plate tectonics, let's look at the phenomenon historically—at how it was discovered and some early thoughts, guesses, and hypotheses concerning this most remarkable reversal. The first suspicion that north-south polarity might have been unpredictably unstable came as early as 1906 from the labs of Professor Brunhe of France and a little later from the lab of Professor Matuyama of Japan. In the 1920's, while routinely examining the magnetic properties of a sequence of layered volcanic rocks, Matuyama found that while the N-S directional line was fairly consistent in all of the layers, some of the rocks indicated a *north* pole in one direction, and others indicated a *south* pole in the *same* direction. Professor Matuyama asserted that the earth's magnetic field had reversed itself at times in the past. This assertion met with much skepticism from the geologic community when he published his hypothesis. However, in the 1950's, when other researchers observed the same phenomena in volcanic rocks in Iceland, the United States, France and in other places, the idea became more believable and gained considerable respect. During the 1950's, however, scientists had not yet related reversal of the earth's magnetic polarity to continental drift. That part of the story came later.

One reason for a lack of enthusiasm for worldwide polarity reversals was that during the 1950's several researchers suggested the possibility that the north-seeking pole of an individual rock might spontaneously reverse to become south-seeking, a phenomenon called "self-reversal." We need not go into this in much detail. Suffice it to say that it was thought that the rocks might have some property by which a self-reversal might occur. However, it has been fairly well demonstrated now that, while magnetic self-reversal *is* possible, it can only account for a *very minor* number of instances of reverse magnetism seen in rock samples. Thus, we are left with the alternative that the earth's magnetic polarity *does* reverse periodically, but for what reason and by what mechanism we still are not certain.

Magnetic Ocean Floor Stripes

The most compelling evidence in support of plate tectonics was discovered in the early 1960's in the form of variations in the strength of the earth's magnetic field. These variations were found as a result of routine magnetic studies of the ocean floor, carried out by towing an instrument known as a magnetometer behind either an airplane or a ship across the ocean floor. (A magnetometer measures the strength of the magnetic field at any point on or near the earth's surface.)

We have already discussed the general magnetic field of the earth, probably produced within its core. The strength of this general magnetic field has been measured (in units called gammas). We know, for example, that the strength of the earth's general magnetic field varies from 30,000 gammas near the equator to 60,000 gammas near the poles.

It was found in the routine magnetic studies of the ocean floor that the total strength of the magnetic field in certain places varied significantly from the general magnetic strength. The measured values of magnetic field strength were either higher or lower than the known general value. These discrepancies were labelled **magnetic anomalies.** (An anomaly is defined as a deviation from the common rule.)

As was pointed out in a previous section, the ocean floor is a layered complex of igneous rocks (basalt) overlain by sediment that has settled to the ocean floor. The rocks comprising the ocean floor have a magnetism due to the presence of magnetic minerals in these rocks. Magnetic anomalies are apparently due to the contribution of the built-in magnetism of the crustal rocks. To be identified as an anomaly, the contribution from the rock must be on the order of 100 gammas.

Recall our discussion of the earth's periodic magnetic field reversals. Rocks that form during a period of "normal" polarity have a magnetic field orientation that is in the same direction as the earth's general magnetic field direction as it exists today. Any measurement of the total magnetic field over a rock that was formed during a period of normal polarity will yield a value *higher than* the earth's general magnetic field strength (**positive anomaly**).

By the same token, a rock that forms in a period of reversed polarity will have a magnetic field orientation in the opposite direction to the earth's general magnetic field direction. Any measurement of the total magnetic field strength over a rock that was formed during a period of reversed polarity will yield a value *lower than* the earth's general magnetic field strength (**negative anomaly**).

The absolute value of these anomalies depend upon the strength of the magnetic field of the rock being studied, but for most igneous rocks, the value of the anomaly is in the range of tens to hundreds of gammas.

It has been possible to map these anomalies in order to determine if they are arranged in discernable patterns. Any consistent patterning might shed light on the origin of these anomalies. Lo and behold, as the mapping proceeded, strong and distinctive patterns did, in fact, emerge. Positive and negative anomalies were seen to define linear patterns on the ocean floor—alternating long "stripes" of positive and negative anomalies. What's more, these stripes were seen to parallel another common feature of ocean floors—long, linear submarine volcanic mountain belts. In fact, the sequence of positive and negative stripes on one side of the mountain chain were found to be mirror images of the sequence on the other side of the mountain chain. Figure 6.5 shows the world-wide distribution of magnetic anomalies associated with the volcanic mountain chains. Is this coincidence or is there a connection between linear magnetic anomalies and linear submerged volcanic mountain chains?

In 1963, two British geologists, Fred Vine and Vince Matthews proposed that the measured oceanic floor stripes indicated that bands of basaltic crustal rock had become magnetized during periods of alternating polarity of the earth's magnetic field. They concluded *that the sea floor was upwelling and spreading away from these submarine volcanic mountain chains,* which ap-

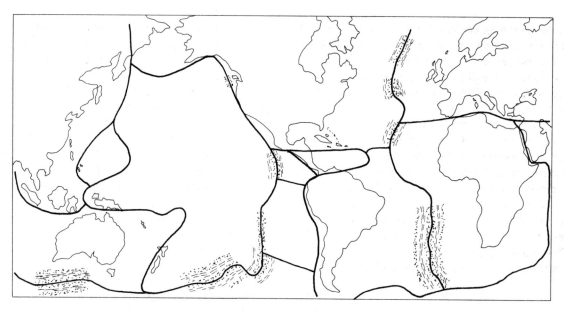

Figure 6.5. World map showing magnetic anomalies associated with spreading-center plate boundaries. Note several mirror images. After M. Nafi Toksoz, "The Subduction of the Lithosphere," *Scientific American,* November 1975, pp. 90–91. Copyright © 1968 by W. H. Freeman and Company, Publishers, San Francisco. Used with permission.

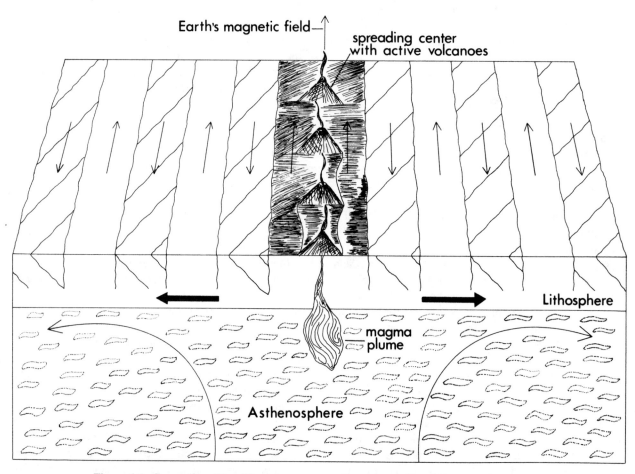

Figure 6.6. Generation of basaltic magma at spreading centers. Note that away from the spreading center, on both sides, the older basaltic rock shows alternating normal and reverse polarity.

pear to be **spreading centers.** Seldom has an idea had such a profound effect on science. This Vine-Matthews hypothesis on sea-floor spreading revolutionized the earth sciences. We will continue throughout this book to come back to this hypothesis and to look at the far-reaching ramifications of it.

For now, however, let's look in detail at how sea-floor spreading works. The basic idea is that basaltic magma is generated rather continuously at these spreading centers. Further, by some mechanism which is still not fully understood, this basaltic crust that is formed at the spreading center, is rather continuously forced or pulled *away from* the spreading center in both perpendicular directions, as shown in Figure 6.6. This idea then would suggest that the age of the crustal basaltic rocks increases away from the spreading centers.

This was an easily tested hypothesis, and it turned out that, in fact, the basaltic crust is rather uniformly older away from the spreading center, in both directions. These data were obtained by means of radiometric dating (see Chapter 3) of rocks sampled systematically away from spreading centers. Thus, it is now generally accepted that these submarine volcanic mountain chains such as the Mid-Atlantic Ridge and the East Pacific Rise are zones along which *new* crustal material is being continuously generated and, furthermore, that these zones are places where movement of crustal material relative to other crustal material is taking place. We are now certain that large-scale movement is continuously taking place and involves not only the continental crust (as Wegner proposed) but *all* crust, continental and oceanic. We still do not know the mechanism by which this movement takes place, but speculation on that problem will be examined in the following chapter.

CHAPTER 7
The Nature of the Crust and Upper Mantle

Plates are apparently continuously in motion, and, because they are in motion, plates interact physically with one another. Plate borders are classified in the following ways:

1. **Divergence**—where plates are moving away from one another.
2. **Convergence**—where plates are moving toward one another.
3. **Transform**—where plates are moving past one another.

No doubt there is no place where any of these three motions are "pure," i.e., divergent plate boundaries almost always have some transform motion associated with them, as do convergent boundaries. Likewise, transform boundaries almost always have either a divergent or convergent component. In this book, plate boundaries will be referred to as either divergent, convergent, or transform, depending on which is the dominant interaction.

The relative movement of plates generates earthquakes in all cases; volcanoes in some cases; and mountain belts in some cases. Chapters 9 and 10 describe those deep processes occurring at each of the three types of plate boundaries. First, we present some introductory descriptions and definitions of the nature of the crust and mantle and of earthquakes, volcanoes and mountain belts.

The Earth's Crust

Figure 7.1 shows diagrammatically the nature of the earth's crust and mantle. One very important feature to note is that the earth's crust varies greatly in thickness. Oceanic crust is uniformly about 7 km thick, except at ridges where it can be a little thicker. Continental crust is considerably thicker, and the thickness is quite variable.

Not only does the earth's crust vary in thickness, its composition is very variable. Continental crust has been sampled for a couple of hundred years. Early descriptions were based primarily on the mineralogical composition of the rocks and texture of these rocks (the size and shape of the mineral grains and the spatial relationships of mineral grains to one another). Later,

chemical analyses were made of these rocks. One conclusion reached is that continental crustal rocks are very heterogeneous, more different than alike. Continents contain all three of the major rock types: igneous, sedimentary and metamorphic and every known type of each of these major rock types. Based on *all* of the data collected, geologists have determined that the continental crust has an average composition, by weight of:

oxygen (O) 46.6%	iron (Fe) 5.0%
silicon (Si) 27.7%	calcium (Ca) 3.6%
aluminum (Al) 8.1%	magnesium (Mg) 2.1%
sodium (Na) 2.8%	potassium (K) 2.6%

(Mason, 1966, p. 48)

(Be aware that this is an *average*; no one rock type has this exact composition, and various rock types have very different compositions.)

Because of its relative inaccessibility, the nature of oceanic crust has not been known for nearly as long, nor has oceanic crust been sampled as extensively as continental crust. However, samples of oceanic crust have been obtained by drilling into the sea floor. The Deep Sea Drilling Project, an ongoing project funded by the National Science Foundation, has provided the vast majority of the data on the composition of oceanic crust on which we base our model of oceanic crust.

Oceanic crust contains really only two components: (1) an upper 1–2 km layer of unconsolidated sediment and (2) a lower 5–8 km basaltic crust. The sediment layer is a surface feature that is probably not related to the origin of oceanic crust. It is "fallout" from the overlying water. We will not discuss this sediment any further in this chapter, but will discuss it in some detail in Chapter 12.

The 5–8 km of oceanic crust underlying the sediment is much more uniform in composition than continental crust. It consists almost entirely of basalt, an igneous rock, with lesser amounts of metamorphosed basalt, amphibolite. The average chemical composition of the basaltic (basalt-like) layer is:

oxygen 45.6%	magnesium 4.3%
silicon 23.0%	calcium 8.4%
aluminum 9.0%	sodium 2.0%
iron 6.7%	potassium 0.1%

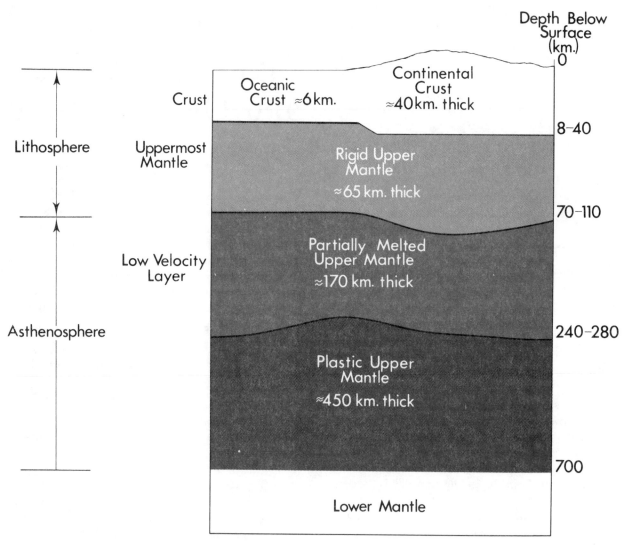

Figure 7.1. Physical characteristics and approximate thicknesses of continental and oceanic crust and mantle. Note that the boundary between the lithosphere and asthenosphere does not correspond to the boundary between crust and mantle.

Two points should be noted. First, most of the oceanic basaltic layer has a composition very near the average given above. Second, the average composition of oceanic crust is quite different from the average composition of continental crust.

Although we have no *direct* evidence of the composition of those very deep parts of the continental and oceanic crust that have not been reached by drilling, seismicity studies have provided us some *indirect* evidence of its composition. Laboratory experiments have shown that mineralogically different kinds of rock transmit seismic waves at different speeds or velocities. Seismic waves move with different velocities through different rocks, speeding up and slowing down depending primarily on the composition and density of the rocks through which they travel. Seismologists can thus infer something about the composition of deep crustal rocks by studying how long it takes seismic waves to move from the earthquake focus to their seismograph stations.

The Earth's Upper Mantle

The earth's mantle has never been *directly* sampled. (We stress the word "directly" because many geologists believe that occasionally pieces of mantle get incorporated in the crust.) Even though we may have no direct evidence of the composition of the mantle, we know that the earth has a mantle beneath the crust, and we have a reasonably good idea of its composition.

The base of the crust (and therefore the top of the mantle) is delineated by a dramatic increase in the velocities at which seismic waves are transmitted, as shown in Figure 7.2. This dramatic change is known as a **velocity discontinuity.** The velocity discontinuity is known as the Mohorovicic Discontinuity (Moho for short) after the Yugoslavian geophysicist who discov-

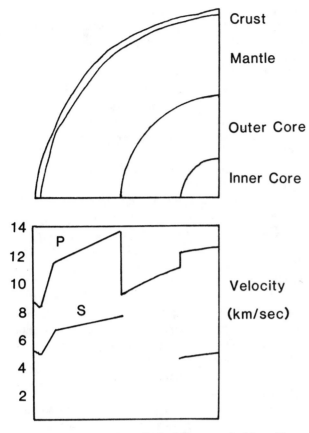

Figure 7.2. Variation of P and S wave velocities with depth.

cally complex system and rarely melt at a single temperature even if held at a given temperature for a long time. Rather, rocks melt over a *range* of temperatures, just as magma solidifies over a range of temperatures. (This was discussed in Chapter 6, in the Section "Rock Magnetism.") Furthermore, laboratory studies show that the first melt that evolves from a rock subjected to high temperatures does *not* generally have the same overall chemical composition as the parent rock. That means that the remaining solid portion of the heated rock must change its overall chemical composition as melting proceeds. This phenomenon is known as **partial melting.** To be certain, if the rock is taken to a high enough temperature, it will totally melt, and the melt produced will be the same composition as that of the rock under investigation. However, total melting requires an ever-increasing temperature and a long period of time. Most melting in the mantle takes place at a constant temperature, lower than the temperature at which total melting of either eclogite or peridotite takes place. The laboratory studies on rock melting, mentioned earlier, verify what you may have already deduced: *Partial melting of peridotite yields a melt, which when allowed to cool, yields the rock basalt.* Partial melting of eclogite yields a melt whose composition is quite different from a basaltic melt. Thus, we think the mantle is essentially peridotite in composition.

Lithosphere and Asthenosphere

We now have what we believe to be a reasonably accurate picture of the chemical makeup of the crust and mantle. What about their physical properties, though? In Chapter 1, we mentioned that the plates behave like *brittle* solids, i.e., they fracture when subjected to pull-apart stresses, and we mentioned that the material beneath the plates behaves more like a *plastic* solid, i.e., one that flows, albeit very slowly, when subjected to the same kinds and magnitudes of stress. Based on what we now know about plates, we believe that the base of the plates is *not* at the base of the crust, as we might initially guess, but that the base of the plates is *within* the upper part of the mantle. Figure 7.1 shows the relationship between crust, mantle and plate nomenclature. Note that the crust and that brittle-behaving part of the uppermost mantle that together comprise the plates are known as the **lithosphere.** The mantle below the lithosphere, that part that behaves as a plastic solid, is known as the **asthenosphere.**

You might suspect that such dramatic changes in physical properties from a brittle lithosphere to a plastic asthenosphere is due to some drastic chemical change. Yet, we believe the contrary, that the lithospheric upper mantle and asthenospheric upper mantle are probably the same peridotitic composition. How can that be? To understand this phenomenon, we must un-

ered it in 1909. We infer that these changes in seismic velocity are due to the seismic waves being transmitted through mineralogically different material. Because seismic waves travel at significantly different velocities in the crust and mantle, we believe that the mantle is significantly different from the crust.

Extensive studies of the velocities at which seismic waves are transmitted in the mantle lead us to believe that the mantle is relatively homogeneous mineralogically. Laboratory studies of seismic-wave velocities in various kinds of rocks, at temperatures and pressures thought to exist in the mantle, constrain us to believe that the mantle is composed of either rock type eclogite or peridotite. In order to choose between these two, we must take into account one other constraint. We believe that the basaltic lava produced at oceanic ridges such as the Mid-Atlantic Ridge and East Pacific Rise is coming from the mantle. So, the mantle must have an appropriate composition to yield basalt.

Eclogite, which is a high-temperature and high-pressure metamorphic equivalent of basalt, might seem to be the obvious choice. Chemically, it is virtually identical to basalt (but mineralogically different due to metamorphism); whereas peridotite is quite different chemically from basalt. Again, however, laboratory studies of the melting behavior of rocks provide us with the key to resolving this question. Rocks are a chemi-

derstand how changes in temperature with increasing depth and changes in pressure with increasing depth affect rock. Refer back to Figure 2.3 to see the relationship between temperature and depth and the relationship between pressure and depth.

When the temperature of a rock reaches the onset of melting, the rock begins to lose its strength, i.e., it becomes weaker and more easily able to flow as a plastic solid. Higher pressure, on the other hand, tends to keep a rock from melting. (Ice behaves differently and melts at lower temperatures when subjected to higher pressure, as, for example, the situation in ice skating.) A rock that melts at one temperature at a given pressure must be raised to a higher temperature to cause melting if the pressure is higher. Thus, we have two counteracting effects, increasing temperature tending to weaken a rock and, at the same time, increasing pressure that tends to strengthen the rock. At depths of 70–110 km, mantle rock begins to weaken considerably. Seismic waves can be seen to travel more slowly in this zone of weakness, known as the **low-velocity zone.** This zone marks the top of the asthenosphere (base of the lithosphere) and extends approximately 150 km within the asthenosphere. Here, the peridotitic asthenosphere is in a partially molten state. Above the low-velocity zone, the temperature is too low for partial melting. Below that zone, even though the temperature is even higher than in the low-velocity zone, the pressures are so great as to prevent partial melting. However, although partial melting is not thought to be occurring below the low-velocity zone, the peridotite still behaves like a plastic solid and is still called the asthenosphere.

The Driving Force for Plate Tectonics

Throughout this book, beginning even in Chapter 1, we discussed the motion of plates but did not explain the "how" or the "why" of the motion. Throughout this chapter, we continue to make the assumption that movement takes place, an assumption believed by most geologists and geophysicists. Whereas there now exists a consensus among earth scientists that motion is taking place, there is still *wide* diversity in what these people believe are the causes.

Let us consider the most widely accepted hypothesis concerning the driving force(s) of plate tectonics: the mantle convection hypothesis. Convection is a heat-transfer process in which heat is transferred from one area to another through mass movement. This is the principal mechanism by which heat is transferred in water being heated by a burner at the bottom of a ves-

sel. In general, the density of a solid or a liquid varies inversely with the temperature of the solid or liquid, i.e., the hotter a material, the lower its density and vice versa. The bottom water heats up by heat transfer from the burner. The warmer bottom water becomes less dense than the colder, over-riding water; the warm, low-density water then rises to the top, displacing the colder water which must sink. The process then begins over again and continues until a temperature equilibrium is attained.

Consider how this mechanism might work in the mantle. First of all, we encounter a fundamental difference between our water example and the mantle: The mantle is not liquid but is composed of hot, solid but plastic rock. It is generally thought that the lower part of the mantle is hotter than the upper part of the mantle (see Figure 2.3). Because the lower mantle rock is hotter, it has a lower density; therefore, this material might have a tendency to rise like the heated water. The problem is that the mantle is *solid* (albeit plastic), and the question we must ask becomes: Can convection take place in a plastic solid, especially one comprised of what we believe to be the composition of the mantle? This is the heart of the controversy concerning the mantle convection driving-force hypothesis.

Believers in the mantle convection hypothesis suggest that the mantle is comprised of a number of **convection cells** in which warm, less dense mantle rises and displaces cooler, more dense mantle, in a continuous process. As the warm, low-density plume of mantle rises, it becomes progressively cooler; eventually, the upward motion must cease. But because this is a continuous process, additional warm, low-density mantle is rising all the time, from the same general source area. In order to make room for the upward rising plume, the earlier material is forced *sideways* from the plume in a manner similar to that in Figure 7.3. Eventually this material moving sideways cools sufficiently for it to start to sink, to replace the material that is rising. Figure 7.3 shows two complete convection cells. The convection cells, if they exist, are three dimensional rather than two dimensional as suggested by Figure 7.3. That is to say that the "spot" where mantle plumes rise is actually a horizontal line of plumes rising, as seen in Figure 7.4. Apparently, the same process does *not* take place in the lithosphere, because the lithosphere is *rigid* and not capable of plastic deformation, and yet it is the lithosphere that shows evidence that convection is occurring beneath it. If convection is taking place in the asthenosphere and *if* the lithosphere and asthenosphere are physically coupled (so that if one moves, the other must also move), then we can explain various plate motions. Where the plume reaches its highest point and

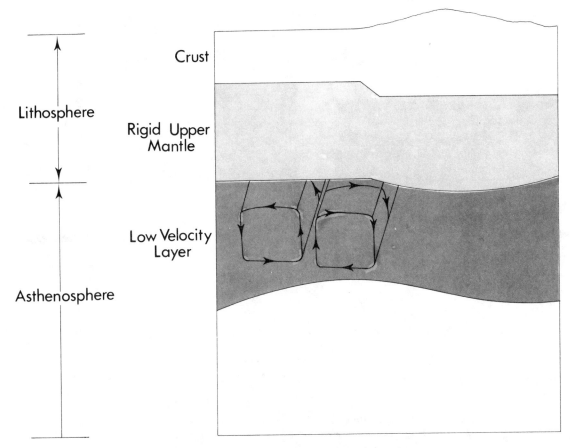

Figure 7.3. Two convection cells in the asthenosphere.

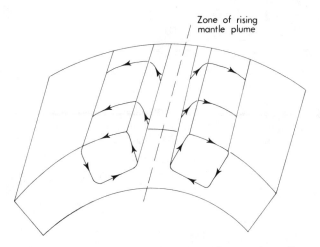

Figure 7.4. Three-dimensional nature of convection cells.

starts to diverge, the lithosphere *above* this region finds itself being dragged in opposite directions (see Figure 7.5) due to the coupling with the asthenosphere. Above the place where the convection cells in the asthenosphere start to descend, the lithosphere in the overlying two cells will converge (see Figure 7.6).

This relatively straightforward process would provide us with the appropriate physiographic features and products *if convection does indeed occur in the asthenosphere*. Many geologists and geophysicists now believe, on physical grounds, that this process is impossible, and they must call upon a different driving-force mechanism.

Another driving-force process that has been suggested is the *sinking slab hypothesis*. The fundamental concept on which this hypothesis is based is that cool lithosphere is more dense than the warm lithosphere. If this is so, then the pull of gravity would be greater on the denser material and could initiate downward movement of the cool lithospheric slab.

We have not attempted to describe any of the above processes quantitatively. Both may be plausible. In this book, we shall consider the driving forces on plates to be some combination of convection and slab-pull forces.

Reference Cited

Mason, B., 1966, Principles of geochemistry: John Wiley and Sons, Inc., New York, p. 48.

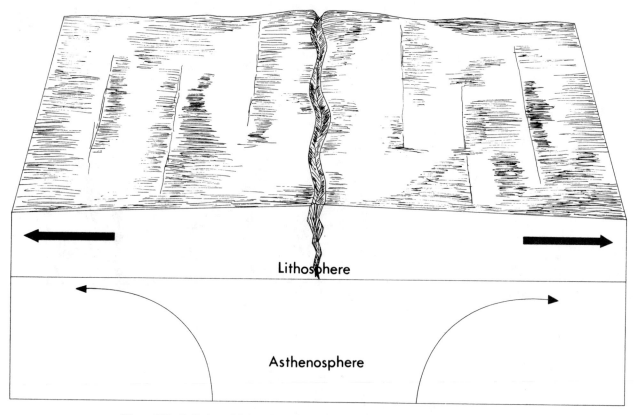

Figure 7.5. Splitting of lithosphere due to drag in opposite direction. Lithosphere is responding to divergence of convection cell in the asthenosphere.

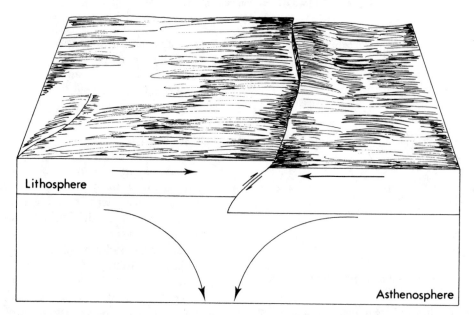

Figure 7.6. Convergence of lithosphere in response to convergence of convection cells in the asthenosphere.

CHAPTER 8
Deformation of Lithosphere

Because plates are always in motion, and plates are constantly interacting with one another, deformation of the lithospheric plates is particularly noticeable at plate boundaries. In this chapter, we want to introduce you to the various types of deformation that take place in the lithosphere. To do this, we will again draw upon data generated by laboratory experiments in rock mechanics.

We have already noted that, in general, the lithosphere behaves like a brittle solid, i.e., it tends to fracture when stressed; whereas the asthenosphere behaves more like a plastic solid, i.e., it flows when stressed. Even though these generalizations are basically valid, it has been shown that deformation of the lithosphere is considerably more complex. Depending on the nature of the stresses that are causing the deformation, the lithosphere can behave like either a brittle or plastic solid. The parameters that govern how lithosphere will deform include: (1) temperature, (2) confining pressure, and (3) rate of stress application.

Stress and Strain

Before examining the effects of each of these parameters, let's first look at how rocks deform, in general. To do this, we must define two terms: stress and strain. **Stress** is the force that is applied to a material that tends to change its form or volume; actually, it is measured in pounds per square inch (psi) or dynes per square centimeter. It is defined as force divided by surface area on which the force is acting. This is an important point. Both the force and the area that a given force is acting on are equally important. Consider putting a one-pound weight on the head of a straight pin and what this one-pound weight, focused on the pin point feels like on your arm. Next, consider a one-pound sheet of plastic 3″ by 3″ and how it would feel resting on your arm. Both are exerting one pound of force, but because of how the forces are distributed, the effect on your arm would be quite different. **Strain** is the amount of change that takes place in a material that is being stressed. If we push hard enough on a cylinder of rock (stress), the cylinder will become shorter (strain). When stressed, rocks behave in three different ways. At low stresses, the rock will behave **elastically**; if it becomes shorter when stressed, it will resume its original shape when the stress is removed, like a cylinder made of rub-

ber. This elastic behavior of rocks can be demonstrated over a range of stresses. Furthermore, if we graph stress *vs.* strain, we see a straight-line relationship. If we double the stress, the strain will also be doubled. This region of elastic deformation is labelled "A" in Figure 8.1. If a rock is subjected to too high a stress, it will

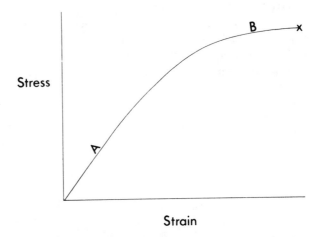

Figure 8.1. Relationship between stress and strain. Region marked "A" corresponds to elastic deformation, "B" corresponds to plastic deformation, "X" corresponds to the fracture point.

deform permanently; i.e., it will not resume its original shape even when the stress is removed. This type of deformation is known as **plastic** deformation. A rock will behave plastically over a range of stresses. In Figure 8.1, the region of plastic deformation is labelled "B." Note that the stress-strain relationship for plastic deformation is not a straight line; a small increase in stress will induce a large increase in strain. Finally, if the stress is great enough the rock will **fracture** (behave as a brittle solid). Point "X" in Figure 8.1 is the fracture point.

Now, let's consider the effects of temperature. A rock that behaves like that shown in Figure 8.1 at one temperature will not behave the same way at different temperatures. In general, as a rock is heated up, it deforms plastically at lower stresses; however, it can withstand even higher stresses before it fractures. This is shown graphically in Figure 8.2, a typical stress-strain diagram for a rock at two temperatures.

Consider next the effect of depth. At greater depths, a rock is subjected to increasingly higher pressures due

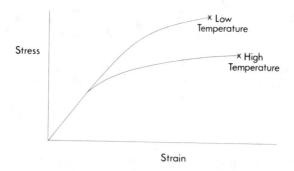

Figure 8.2. Effect of temperature on stress-strain curve.

to the weight of all the rocks overlying it. This pressure is known as the **confining pressure.** Laboratory studies demonstrated that, if a given rock behaves like the ideal rock in Figure 8.1 at one confining pressure, it will not behave the same way at different confining pressures. In general, as the confining pressure increases with depth, the rock can withstand higher stress before undergoing plastic deformation, and it can withstand higher stresses before it fractures. Figure 8.3 is a typical stress-strain diagram for rock at two confining pressures.

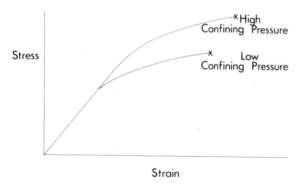

Figure 8.3. Effect of confining pressure on stress-strain curve.

Finally, consider the effect of the rate of stress application. In general, a rock that behaves as a plastic solid when subjected to slowly applied stress may rupture at the same stress if stress is applied rapidly. This is analogous to the behavior of "silly putty." If a stress is applied slowly, the silly putty will flow. If, however, the silly putty is struck with a hammer, even with the same stress, it will fracture. Try it, if you haven't already!

Now, let's try to put all of this information together to see if we can understand the deformation of lithospheric plates. Near the surface, lithospheric material behaves like a brittle solid. Because both temperatures and pressures are low, the region of plastic deformation is very small. In other words, if rocks are stressed near the surface, they behave elastically up to a point; then, if the stress is increased, they fracture without undergoing a significant amount of plastic deformation. At progressively greater depths, when both temperature and pressure are increasing, the same rock will undergo plastic deformation over a large range of stresses, and fracture doesn't take place until even higher stresses are felt.

Faulting

This brings us to the features that are formed when rocks are stressed: faults and folds. A **fault** is a fracture along which movement has taken place. As you can probably deduce from the above discussion, faults tend to occur near the surface at moderate rates of stress application. Faults are a less common form of deformation at depth unless the rates of stress application are fast.

If a brittle solid is subjected to tensional (pull-apart) stresses, until it fractures, the fractures will tend to be perpendicular to the direction of the tension. Figure 8.4

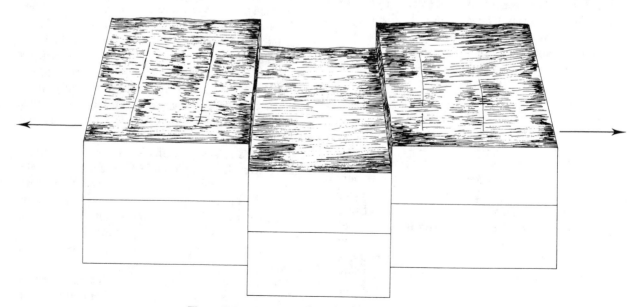

Figure 8.4. Normal fault produced by tensional stress.

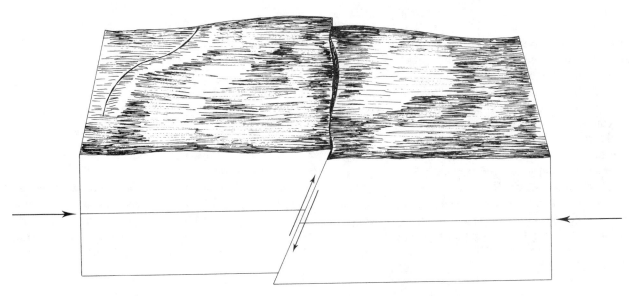

Figure 8.5. Thrust fault produced by compressional stress.

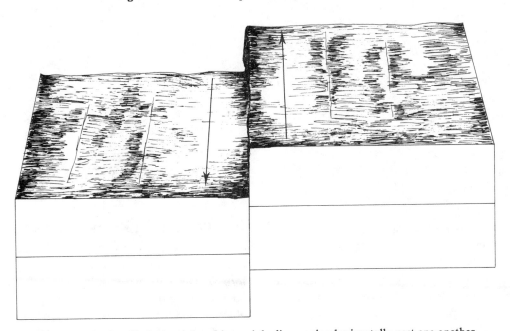

Figure 8.6. Strike-slip fault produced by rock bodies moving horizontally past one another.

shows how a large rock body probably will behave after undergoing tensional stresses. The plane of the fault is predominantly vertical, and it can be seen that most of the movement along the fault is vertical. This is known as a **normal fault.**

Consider next a brittle solid that is being squeezed, or in technical terms, subjected to compressional stress. This material will also fracture, and Figure 8.5 shows how a rock body would fracture into blocks after undergoing compressional stresses. Note that one block has been pushed or thrust up and over the other block. The fault plane is at an acute angle to the direction of the stress, and the movement along the **thrust fault** plane has both vertical and horizontal components. In the case of large rock masses, the fault will often be characterized by pulverized rock material due to the fact that one body is being pushed over another.

Finally, faults can also have predominantly horizontal movement; in this case, rock bodies are simply being pushed past one another. The plane of the fault is essentially vertical, and the fault is known as a **strike-slip** fault. Figure 8.6 shows a strike-slip fault. We will see in Chapter 9 that the faulting that takes place at transform boundaries is a special type of strike-slip fault.

Folding

Whereas the fracturing of rock yields faults, the flowage or plastic deformation of rocks yields **folds.** Again, you can probably deduce that folding takes place near the surface only if the rate of stress application is very slow. At depth, where increased pressure and temperature bring about a larger stress range for plastic deformation, folding will take place at even moderately rapid rates of stress application. In general, then, if near-surface deformation tends to yield faults, deformation at depth is characterized by folding. We can, of course, turn this around when observing deformed rocks in the field. A predominance of faulting indicates the rock body in question probably was near the surface when the stresses were applied. Likewise, a rock body characterized predominantly by folding probably was buried when the deforming stresses were applied. Be careful about the problem of *rate of stress* application, however, when making such generalizations.

Folds are simply bends in rock bodies. Geologists have applied different names to folds which bend upward, **anticlines,** and folds which bend downward, **synclines.** Figure 8.7 shows a series of synclines and an-

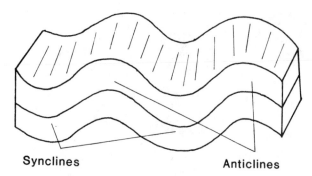

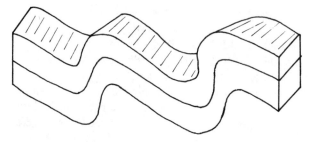

Synclines   Anticlines

Figure 8.7. Anticlines and synclines.

ticlines. Folds can range in complexity from very simple to very complex, depending on the strength of the deformation to which they have been subjected. With very strong deformation, rock bodies can be distorted into **overturned folds,** as seen in Figure 8.8, or, in ex-

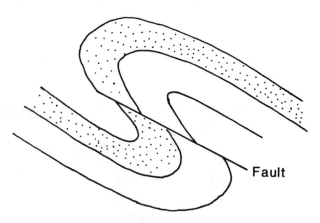

Figure 8.8. Overturned folds.

treme cases, into **refolded folds** as seen in Figure 8.9. These types of folds are very common in metamorphic rocks of intensely folded mountain belts.

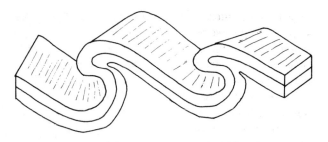

Figure 8.9. Refolded folds.

Some rocks show evidence of being both folded and faulted. Figure 8.10 shows one such example. In this case, the rock responded to deformation first by folding; however, the deforming stresses overcame the strength of the rock, and faulting occurred.

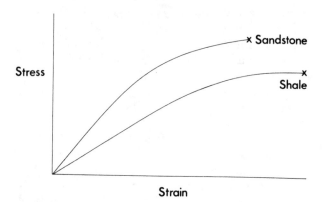

— Fault

Figure 8.10. Fault produced after rock was folded.

Effects of Stress on Different Rock Types

Finally, one must be careful when attempting to determine the conditions under which a rock was deformed. Not only must we keep in mind a complex interplay of temperature, confining pressure, and rate of stress effects on the deformation of rocks, we must also be aware that different rocks behave differently under the same conditions (See Figure 8.11). For ex-

Stress

×Sandstone
×Shale

Strain

Figure 8.11. Stress-strain curves for typical sandstone and shale under same conditions of temperature and confining pressure.

48

ample, in an interlayered sequence of sandstones and shales, which have been deformed under the same total stress, temperature, pressure and stress-rate conditions, it is quite common to see the sandstone layers deformed into folds (plastic deformation) where the shales show evidence of folding (plastic) and slippage along the shale layers (microfracturing).

In the next two chapters, we will discuss what kinds of deformations take place with particular kinds of plate interactions.

CHAPTER 9
Divergent and Transform Boundaries

In this chapter, we will examine both divergent (pull apart) and transform (sliding) boundaries, because in almost all cases, these are "coupled" features—where you find one, you almost always find the other. This is apparently not coincidental—the two types of boundaries are now thought to be related to one another. But how?

Divergent Boundaries

The divergent boundary, or **spreading center boundary,** is that boundary at which plates are moving away from one another. In this chapter, we will consider the physiographic features, magma generation, faulting and earthquake activity associated with this type of boundary, examining modern divergent boundaries and ancient counterparts.

Divergence can take place either in continental lithosphere or oceanic lithosphere and is the result of mantle forces pulling the overlying lithosphere apart. To understand the processes involved and the products derived therefrom, let's pay particular attention to two areas—the Mid-Atlantic Ridge and the East African Rift Valley, one an example of oceanic lithosphere divergence, the other an example of continental lithosphere divergence.

Divergent boundaries are particularly important because it is here that new crust is continually being generated. To understand how new crust is generated, we must first look at what is happening. By whatever cause, forces in the mantle are pulling brittle lithospheric material apart. The lithosphere, then, is experiencing tensional stresses. Recall in Chapter 8, that when tension is applied to brittle material, the material fractures, roughly perpendicular to the direction that the stresses are being applied. Tension applied to lithosphere yields normal faults, with roughly vertical fault planes and predominantly vertical movement along the faults. In both cases of lithospheric divergence, the result is long, linear rift valleys. Figure 9.1 shows a topographic profile across the Mid-Atlantic Ridge. A profile across the East African Rift Valley would be similar but located above sea level.

Note also that in Figure 9.1, the surface on both sides immediately adjacent to the rift valleys is high relative to nonrifted areas. Careful examination of these high areas show them to be sites where magma is rising to the surface. In most cases, the magma is **basalt.** This basalt is new crustal material, and evidence suggests that it is being derived from the mantle. But what is the process by which this basalt is derived? To answer that, we must recall the effects of both increasing tem-

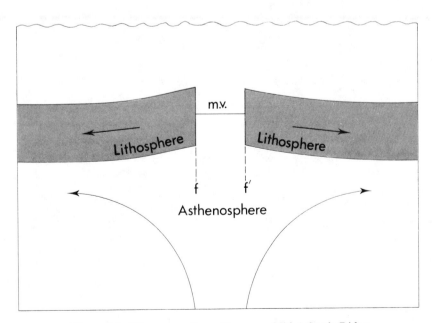

Figure 9.1. Diagrammatic profile across Mid-Atlantic Ridge.

51

Figure 9.2. Close-up view of Mid-Atlantic Ridge system. Fracture zones are roughly perpendicular to the rifted area.

perature and pressure on the process of melting (see p. 42).

The zone or area within the mantle at which melting is taking place is much hotter than necessary to begin to melt the peridotite if it were at the surface; but it is not molten, except below spreading centers. The reason is that the pressure is very high in those areas due to the weight of all of the overlying rock. Increased pressure, recall, keeps the rock from melting. Thus, we have two counteracting factors at depth—increasing temperature that brings on melting and increasing pressure that inhibits melting. Even at high pressure, rock will melt, but the melting temperature is much greater than the melting temperature of the same material at the surface, where pressures are much lower. When lithospheric rifting takes place, pressure is reduced at depth. As overlying material is parted, the pressure of overlying material is reduced. When the pressure drops, the mantle begins to melt. As we noted in Chapter 7, the onset of melting of chemically complex rock material yields a melt different in composition from the rock being melted. In the case of peridotite, partial melting yields basalt. *Thus basalt, new crustal material, is generated at divergent boundaries by partial melting of mantle peridotite.* This generative process is occurring at both oceanic and continental lithospheric divergent boundaries. The only difference is that the basalt that is generated under continental lithosphere tends to react chemically with continental crust on its upward ascent. This results in a magma of slightly different chemical composition at the surface compared with the composition of magma extruded through oceanic crust. Oceanic crust is basalt extruded earlier and does not react chemically with the newly generated basaltic magma.

Many geologists believe that the East Africa Rift Valley, the Red Sea and Atlantic Ocean represent different stages in the divergence or rifting of continental masses. The East Africa Rift Valley is thought to be an example of a very young continental rift. The continental crust is arched upward and block-faulted, with newly generated magma filling the rift-fracture zones. No ocean is present, rather a long, block-faulted rift valley. If this process were to continue so that the continental masses were separated, the newly generated magma in the central part of the rift valley could become new oceanic crust, if the oceans invaded that area. This is apparently the case in the Red Sea where Africa and the Arabian Peninsula have split apart. Continued rifting would yield a wide ocean with a mid-ocean volcanic mountain-rift valley system, a system analogous to the present Atlantic Ocean with its Mid-Atlantic Ridge.

Transform Boundaries

In Chapter 1, we said that the Mid-Atlantic Ridge was a continuous system over thousands of miles. If we look closely at any small part of the Mid-Atlantic Ridge or the East Pacific Rise or any spreading center, we can see that, really, this feature is rarely unbroken for any great length. Rather, spreading centers appear to be fractured into relatively short segments of the system as seen in Figure 9.2, a view of part of the Mid-Atlantic Ridge system. Roughly perpendicular to the rift valley and located between segments are fracture zones, whose origins are not well understood. These fracture zones are very obvious in Figure 9.2. Those parts of the fracture zones between spreading center segments are transform boundaries where plates are moving past one another with *little if any convergence or divergence.* The relationship between divergent and transform boundaries and fracture zones mystified geologists for a long time, and the unraveling of the mystery represents a beautiful example of how the plate tectonic theory continues to evolve.

When geologists first started looking closely at segmented spreading centers, it was thought that the segments were originally continuous (Figure 9.3a) and that segmentation of the system (Figure 9.3b) came as a result of strike-slip faulting—horizontal movement along essentially vertical fault planes (fracture zones).

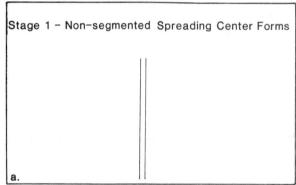

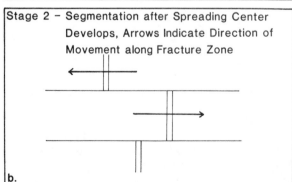

Figure 9.3. Early concept of the development of segmented spreading center. In "stage 1" the spreading center forms; in "stage 2" the spreading center is offset by later strike-slip faulting along fracture zones.

If this were the case, one might expect faulting and resultant earthquakes all along the fracture. Furthermore, the relative movement along the fault should be as noted in Figure 9.3b.

Careful investigations of earthquake epicenters along fracture zones and determination of movement along these faults have yielded a very different picture of the spreading-center–fracture-zone–transform-boundary system. Let's look diagrammatically at a typical piece of segmented spreading center in Figure 9.4. Now, if the first model we described was correct, there should be equal probability of an earthquake to occur *anywhere* along either fracture zone. But we see essentially *no* earthquakes caused by the relative movement of blocks A and C, B and D, D and F, or C and E. Next, according to the first model, any earthquake along FZ_1 (Fracture Zone 1) should indicate movement of blocks A or B moving to the west relative to movement of blocks C and D which would be to the east. Any earthquake along FZ_2 should indicate relative movement of blocks C and D to the east and blocks E and F to the west.

The actual locations of earthquakes and determinations of relative movement could not be explained by the early model. It was found that earthquake epicenters were essentially restricted to those part of the fracture zones *between the segments of the spreading center* (those areas marked "x" in Figure 9.4). Next, it was

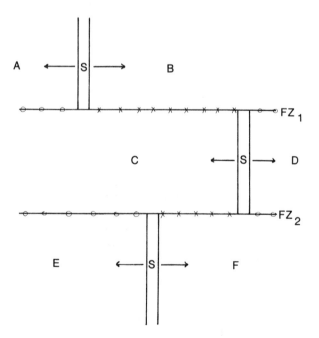

Figure 9.4. Detailed view of segmented spreading center showing location of earthquakes and direction of movement of faults.

found that the earthquakes along FZ_1 were caused by relative movement of blocks B and C such that block B is moving to the *west* relative to block C, moving to the *east*. Earthquakes along FZ_2 were caused by relative movements of blocks C and F such that C was moving to the east relative to block F, moving to the west.

It became apparent that a new model was needed to explain these phenomena. The model that evolved begins with fracture zones as dormant faults on the ocean floor (Figure 9.5a). When the spreading center begins

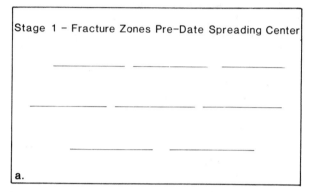

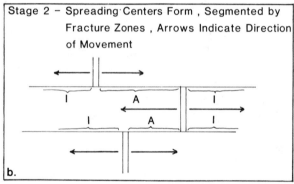

Figure 9.5. Currently accepted concept of the origins of segmented spreading center. In "stage 1" the fracture zones represent dormant faults that pre-date spreading center; in "stage 2" the spreading center develops in segments offset by fracture zones.

to develop, it *actually develops as a segmented spreading center* with each segment bounded by pre-existing fracture zones (Figure 9.5b). Relative movements of blocks on either side of a segment of spreading center are *away from* the spreading center. Thus, in Figure 9.4, block A moves west and block B east; block C west and block D east; block E west and block F east. Therefore, along FZ_1, blocks A and C are moving in the same direction, and blocks B and D are moving in the same direction. There is no relative movement between A and C, or between B and D, so there can be no earthquakes along the fracture zone between A and C, nor between B and D. Blocks B and C, however, are moving in the opposite directions (B to the east, C to the west).

Therefore, this is the part of the fracture zone where earthquakes occur. Likewise, along FZ_2, earthquakes can only occur between blocks C and F.

Thus, it is now believed that the creation of a spreading center revitalizes parts of these dormant faults known as fracture zones. The revitalized parts of the fracture zones become known as transform faults.

Perhaps the best known transform fault in the United States is the San Andreas Fault of California. Figure 9.6 shows the relationship between the North American plate, the Pacific plate, the San Andreas Fault, and the East Pacific Ridge (a spreading center). The areas marked EPR are segments of the East Pacific Rise. The San Andreas Fault can be seen as the transform (active) fault zone of a pre-existing fracture zone. Movement along the San Andreas Fault is such that the Pacific plate, including the southwestern part of California, is moving to the northwest, relative to the North American plate.

Transform plate boundaries then can be seen as a special type of strike-slip faulting. They are characterized by much earthquake activity. Because no convergence or divergence is taking place, no magma is generated along transform boundaries. The physiographic feature associated with transform boundaries include off-sets of natural and manmade features, such as can be seen in Figure 9.6, but no active mountains (either folded or volcanic). In fact, many transform boundaries that exist above sea level are marked by valleys. The faulting pulverizes most rock in the fault zone; this pulverized rock is then easier to erode, frequently leaving long narrow valleys.

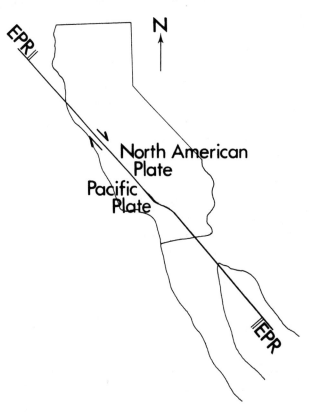

Figure 9.6. San Andreas Fault.

CHAPTER 10
Convergent Boundaries

By definition, convergent boundaries are those at which lithospheric plates are colliding with one another. It is at convergent boundaries that some of our most spectacular geologic phenomena occur: long, linear folded mountain belts such as the Appalachians, Alps and Himalayas; volcanoes, such as Mt. St. Helens, the Andes, Japanese Islands; and numerous earthquakes. Also associated with certain convergent boundaries are the deep arc-shaped trenches found on some deep ocean floors.

The types of phenomena produced at convergent boundaries depend to a very large degree on the type of lithospheric material in collision. We can identify three different types of collisions: (1) those in which two oceanic lithospheric plates are colliding, (2) those in which oceanic lithosphere is colliding with continental lithosphere; (3) and those in which two continental lithospheric plates are colliding. Each generates its own set of tectonic phenomena, and we will discuss these separately.

Ocean-Ocean Collisions

When two oceanic lithospheric plates collide, one is forced under the other. It does not seem to matter which is forced under and which one overrides. Nor is it predictable which plate will move which way. This process of one plate being forced under another is known as subduction, and the process of subduction leads to the tectonic phenomena characteristic of this type of plate interaction—deep ocean trenches, thrust faults and their associated earthquakes, andesitic volcanoes that form island arcs and a particular kind of metamorphism known as **blueschist** metamorphism. Figure 10.1 diagrammatically shows the relation of all of these features to one another at a typical ocean-ocean collision boundary.

Let us first examine the trenches, those typically very long, deep and narrow zones that mark the line along which the subduction plate is forced under the other plate. Recall from Chapter 4 that it has been known since the 1950's that trenches were an area noted for low gravity values and low heat-flow values. Until the

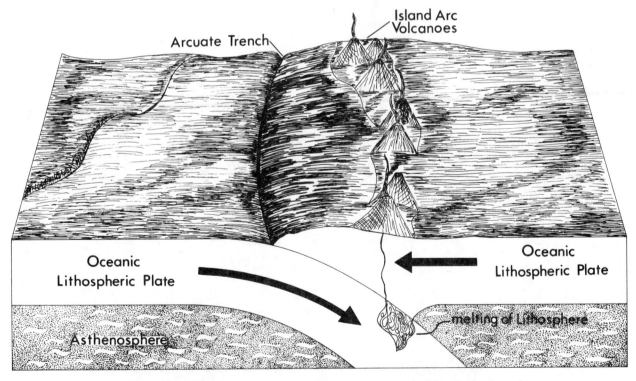

Figure 10.1. Block diagram showing collision of two oceanic lithospheric plates.

dynamic model of plate tectonics came along, the static model of the earth's crust could not explain the geophysical phenomena. With the dynamic plate tectonics model, however, both can be explained.

Low gravity readings over the trenches mean that the mass of rock directly under the point where the gravity reading was taken is less than the mass in other adjacent areas. If the earth's lithosphere was static, this trench area would float upward due to isostasy, eliminating both the trench as a physiographic feature, and the low gravity reading. Low gravity readings in the ocean thus imply that the lithosphere is *not* static—that there must be some force keeping the trench area deep. The process of subduction, where part of the oceanic lithosphere is continually pushed down into the asthenosphere could be that force. And of course, the process of subduction accounts for many of the features associated with trenches—earthquakes with variable depth foci, and andesitic volcanism.

With our concept of subduction of oceanic lithosphere at convergent boundaries, we have a model in which oceanic lithosphere is constantly being forced under other oceanic lithosphere. What happens to the subducted oceanic lithosphere? It descends into the asthenosphere at a relatively fast rate (1–5 cm/yr). This cold, downmoving slab of lithosphere is moving into hotter regions as it descends. Thus, it would stand to reason that the slab would be heating up as it descends. However, and this is a very important point, if the cold

slab is moving down faster than it can heat up (after all, it takes time for an object to heat up even when the surrounding temperature is raised), then the temperature gradient beneath trenches should be such that the temperature at depth increases less rapidly than if the lithosphere were static. If so, it would stand to reason that the amount of heat that is escaping to the surface (the heat flow) is lower over trenches than over surrounding static deep ocean floor.

Consider next the movement associated with subduction. Remember that these are collisional boundaries and that one lithospheric slab is being forced under the other. Movement probably is not continuous because of the tremendous friction between these two slabs. Movement will most likely be sporadic and will only occur when the forces responsible for the convergence overcome the frictional resistance. Sporadic movement such as this leads to faulting, and trenches have long been known to be associated with moderate to severe earthquakes. The faulting associated with oceanic lithosphere convergence is almost exclusively *thrust faulting* (see Figure 8.5), resulting in both horizontal and vertical displacement.

The distribution of earthquake epicenters associated with trenches, as seen in Figure 10.2, has been known for a long time. It has long been noted that shallow-focus earthquakes have their epicenters very near the trenches. Earthquakes whose foci are at moderate depths generally have epicenters located in a band par-

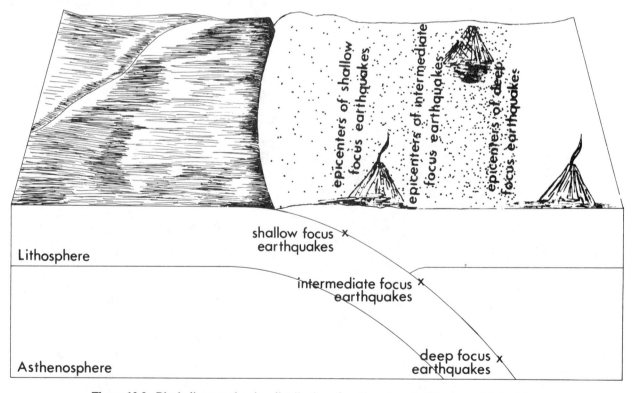

Figure 10.2. Block diagram showing distribution of earthquake epicenters associated with trench.

allel to the trench but shifted laterally away from the trenches. Earthquakes having the deepest foci have epicenters also in a band parallel to the trench but shifted even further from the trench. This distribution suggests that the major fault plane associated with these earthquakes is inclined to the surface. Prior to the development of the plate tectonics model, we had no real understanding of this inclined fault plane beneath trenches. With the plate tectonics model, in which trenches are associated with convergent boundaries, we can now see that the fault plane along which these earthquakes are generated is the boundary between the subducted plate and the overriding plate. The sporadic nature of the earthquake is due to sporadic, noncontinuous subduction in which movement takes place only when frictional resistance is overcome.

What happens to the subducted plate as it descends into the asthenosphere? To be sure, the slab heats up, though slowly, the deeper it goes. It is also subjected to moderately high pressures, suggesting that it may undergo metamorphism. The metamorphism of basalt, the principal component of oceanic crust, leads to the formation of eclogite, a high temperature and high-pressure metamorphic rock having essentially the same overall chemical composition but with a different suite of minerals than basalt. As the subducted slab descends even further into progressively higher temperature regions, the basalt or eclogite will ultimately begin to undergo partial melting. This is the same process by which basaltic magma was generated from the upper mantle peridotite, *except* that pressure reduction was responsible for the peridotite-to-basalt partial melting. Beneath the trenches, partial melting takes place due to increased temperature.

It has long been known that volcanic activity associated with trenches produces, in general, **andesitic** magma which is both chemically and mineralogically different from the basaltic magma associated with mid-ocean ridges. Prior to the development of the plate tectonics model, we did not understand why two such different magmas were produced, in the kind of pattern found on the ocean floor. Now we can put even this into the plate tectonics framework. Laboratory studies have shown that when a rock having the composition of basalt or eclogite is subjected to high temperatures and pressures, the first partial melt will have a composition very close to that of andesite (more silica, less iron and magnesium than the starting material). Therefore, when a basaltic oceanic lithospheric slab is subducted, it heats up and begins to undergo partial melting to produce andesitic magma. This hot magma rises, because it is less dense than the overlying rock, and, eventually, surfaces in the form of andesitic volcanoes. Some of the magma does not make it to the crust surface, but rather cools beneath the surface, forming the plutonic equivalent of andesite, known as **diorite.** Andesitic vol-

canoes form arc-shaped patterns known as island arcs (henceforth, we will refer to the arc as a **volcanic-plutonic arc**) that always appear on the inside curve of the trenches associated with them, more or less parallel to the epicenters of the moderate and deep-focus earthquakes. This is the origin of such island arcs as the Japanese Islands and the Marianas Islands in the Pacific and the Virgin Islands of the Caribbean.

What ultimately happens to the subducted slab, after undergoing partial melting? We do not know whether it maintains its chemical and mineralogical integrity or whether it becomes chemically consumed back into the asthenosphere. But we do have some *indirect* evidence that suggests that it is somehow consumed within the asthenosphere. The deepest focus earthquake associated with trenches is at about 700 kilometers. Above that depth, apparently, the subducted slab maintains its integrity (behaving as a brittle slab) and moves forcibly with respect to the asthenosphere. Below that depth, there is no brittle deformation. The lack of brittle deformation could be explained in several ways. The two most plausible seem to be either: (1) that the slab (maintaining its chemical integrity) no longer behaves *physically* like a brittle solid; or (2) that the slab at that depth is chemically consumed back into the asthenosphere with the slab becoming indistinguishable from surrounding asthenosphere. Geologists and geophysicists are still split when it comes to discussing what happens to those slabs below 700 km.

What happens to that thin veneer of sediment on the basaltic oceanic crust at places of convergence? This material along with volcanic debris from the associated island arc, and some of the uppermost part of the basaltic crust, is forced down deep into the lithosphere by subduction and is subjected to a special type of metamorphism known as **blueschist** metamorphism. This type of metamorphism takes place under conditions of high pressure and relatively low temperature. The high pressure is due to the weight of overlying rock and to the compressional nature of the convergent boundary (squeezing the material together). The temperatures are not as high as we might have expected because of the subduction of one slab and the resulting lower temperature gradient beneath trenches, as noted earlier. Blueschist metamorphism yields a distinctive metamorphic rock, often containing a very distinctive blue mineral **glaucophane.** This blueschist metamorphic material tends to get welded onto the front of the volcanic-plutonic arc, between the arc and the trench.

Finally, the volcanic-plutonic arc, after its inception, is still in an area undergoing compression. Much of the material is forced down into the lithosphere where pressures and temperatures are higher than at the surface. Because of this, it undergoes a high-temperature and high-pressure type of metamorphism that more typically is associated with continent-continent type

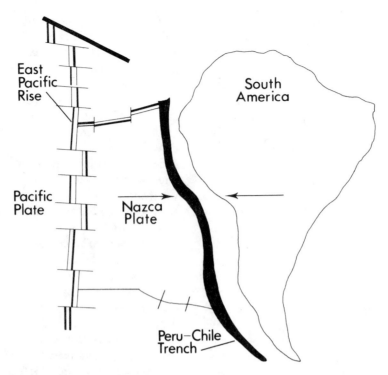

Figure 10.3. Location and movement of Nazca plate relative to Pacific and South American plates.

collisions, producing the more typical metamorphosed igneous and sedimentary rocks, formed by progressively higher pressures and temperatures.

Ocean-Continent Collisions

The second type of collision we will examine is that which involves the collision of oceanic lithosphere with continental lithosphere. One example is the west coast of South America where the South American plate (continental) is in collision with the Nazca plate (oceanic). The Nazca plate is a small plate that sits on the east side of the East Pacific Rise, a divergent boundary separating it from the Pacific plate (see Figure 10.3). The Nazca plate is moving eastward, away from the East Pacific Rise, and the South American plate is moving westward away from the Mid-Atlantic Ridge.

The features associated with this type of collision are very similar to those produced by ocean-ocean collisions. We will examine the processes involved and the features produced, emphasizing those processes and features that differ from the ocean-ocean collisional boundary.

One of the most obvious features associated with the Nazca-South American plate collision is the Peru-Chile Trench lying offshore along the entire west coast of South America. This trench and its associated earthquakes originate in much the same way as ocean trenches and earthquakes in our previous discussion. When oceanic lithosphere is in collision with continen-

tal lithosphere, the oceanic lithosphere is subducted beneath the continental lithosphere in much the same way as in an ocean-ocean collision, except more predictably: the ocean plate is *always* subducted beneath the continental plate. Thus, the Peru-Chile Trench is the result of the Nazca plate being subducted under the South American plate. The distribution of earthquake epicenters in this area is therefore very predictable and explainable. Shallow-focus earthquakes have epicenters offshore and near-shore; progressively deeper focus earthquakes have epicenters progressively further eastward or inland. Seismic studies verify that most of the faulting that generates the earthquakes is *thrust faulting,* again predictable on the basis of a model of a subducting, eastward moving Nazca plate and an overriding westward-moving South American plate.

The second prominent feature at this particular plate boundary is the Andes Mountain chain, that active volcanic chain running the length of inland South America. Again, predictably, the volcanism is principally andesitic (named after the Andes Mountains) due to the partial melting of the Nazca plate (oceanic-basaltic) under South America. Again, much of the magma surfaces to form the Andes volcanic peaks, but an even greater amount of the magma does not reach the surface and crystallizes within the crust as diorite or granitic plutonic rocks. In some parts of the Andes, erosion has removed much of the overlying volcanic rock (andesite), exposing the plutonic (diorite) portion of this volcanic-plutonic complex.

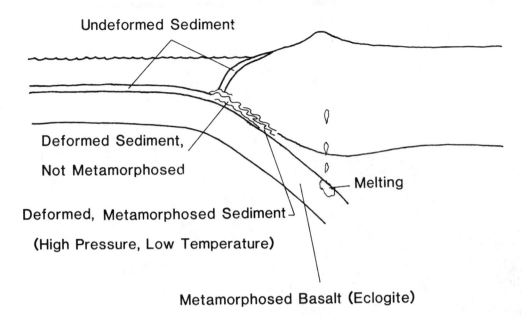

Figure 10.4. Collision of oceanic lithospheric with continental lithospheric plate.

Blueschist metamorphism is even more pronounced in this type of collision than in an ocean-ocean collision, due primarily to the presence of a greater amount of sedimentary rock material to become deformed. Let's examine a typical ocean-continent convergence, as seen in Figure 10.4. Note that the volcanic activity is inland from the offshore trench. The area between the volcanic mountains and the trench is known as the fore-arc basin. This area is continental in nature, in some places above sea level, in other places below sea level. The relative amounts of area above and below sea level depend on the amount of continental shelf involved. On the west coast of South America, there is very little continental shelf. The fore-arc basin is basically an area of sedimentation, and the details of the type of sediment will be left to Chapter 14. Generally, the sediment deposited in the fore-arc basin comes primarily from the mountains, and the total amount of sediment deposited can be very large.

That portion of the fore-arc basin sediment lying nearest the trench becomes involved in the collisional process, along with the sediment on the oceanic floor and the uppermost part of the basaltic oceanic lithosphere. The deep-ocean sediment and upper layer basalt get "scraped off" when the oceanic slab is subducted and becomes "welded on" to the continental margin. Thus, this type of continental margin is said to be growing or accreting. This type of deformation is again characterized by high pressures and low temperatures. The fore-arc basin sediment nearest the trench also gets involved in this blueschist type of deformation and metamorphism. The result is a chaotic mixture of blueschist metamorphosed counterparts of deep-ocean sediment, basalt, and fore-arc basin sediment. Because the total volume of sediment involved is so great, this

"welded on" pile undergoes both brittle and plastic deformation often characterized by overturned, occasionally sheared-off folds.

The sequence of rocks and physiographic features in California, including the Sierra Nevada Mountains, Great Valley and the Coast Ranges, are excellent examples of an ancient ocean-continent collision. The word *ancient* is important here; this collision is no longer taking place. (Remember the present boundary between the North American and Pacific plates is a transform boundary.) The Sierra Nevada Mountain chain is the ancient volcanic-plutonic arc, even though most of the volcanic rocks have weathered and eroded away, exposing the more plutonic *granodiorite* cores and progressively metamorphosed sequences representing the high-temperature, high-pressure metamorphism of the volcanic-plutonic arc. The Great Valley contains a sequence of sedimentary rocks rather easily identifiable as fore-arc basin deposits. The Coast Range, with its Franciscan Formation, is a chaotic sequence of blueschist metamorphic rocks, some identifiable as originally deep-ocean sediments, some as basalt, and some as originally fore-arc basin sediments. All have been intensely deformed into a folded mountain chain that shows evidence of being "welded on" to the previous continent and becoming a part of an accreted continent.

Continent-Continent Collisions

The third type of collision, the continent-continent collision, is responsible for some of the most remarkable and awe-inspiring physiographic features on the earth, the folded mountain belts of the Himalayas, Alps and Appalachians. The Alps and Appalachians are no

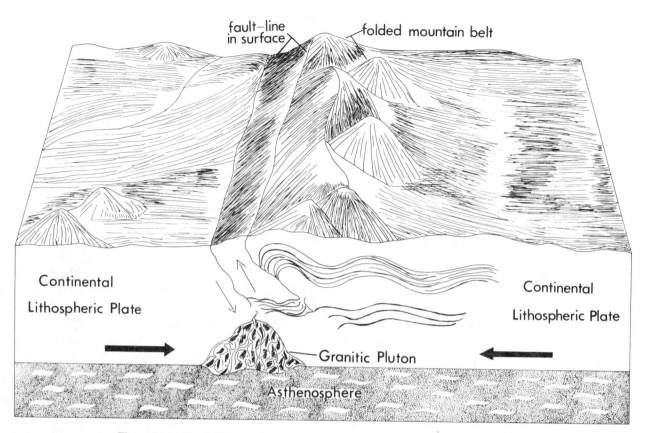

Figure 10.5. Block diagram showing collision of two continental lithospheric plates.

longer in the process of being formed or deformed. The Himalayas, containing Mt. Everest (at 29,028 feet, the highest point on earth) are still undergoing deformation, due to the collision of the Indian and Eurasian plates and serve as our present model for folded mountain belt generation.

To understand continent-continent collisions, it is necessary to understand the nature of continental crust. Recall from Chapter 7 that continental lithosphere is generally thicker and less dense than oceanic lithosphere. When continental masses collide, little or no subduction takes place. (See Figure 10.5.) Perhaps the continental lithosphere is too thick or too light to be forced under other continental material. What does happen is that when collision takes place, the continental material near the collisional boundary, or **suture,** is shortened and thickened by buckling or folding. If subduction does not take place, then crustal shortening must take place; and if crustal shortening takes place, then the shortened material must be elevated high into the atmosphere.

Recall also that continental crust is composed of all kinds of rocks—igneous, metamorphic and sedimentary, all with different deformational properties. Depending on the rate at which the collisions are taking place and the type of rock at or near the suture zone, different structural features are produced. In general, those rocks that are near the surface and near the su-

ture will be *faulted;* whereas those rocks deeper underground and perhaps further from the suture will be *folded,* but it is very clear from detailed studies of folded mountain belts that folding and faulting are frequently intimately associated. Near-surface folding and deep-seated faulting are almost as common as near-surface faulting and deep-seated folding.

Continent-continent collisional boundaries are also characterized by earthquakes, as seen in Figure 5.1. Seismological studies of these earthquakes associated with the faulting in or near the suture zone, indicate that the faulting is principally of the *thrust fault* variety. The thrust faulting is not associated with subduction but fracturing due to high degrees of compressional deformation. Rather than subduction, an important component of the fault movements is the process of **obduction** where one rock mass rides up and over (rather than down and under) another rock mass. (In both subduction and obduction, one plate, or part of a plate, experiences vertical movement. In *subduction,* one plate stays at a constant level while the other is forced downward under it. In *obduction,* one plate stays at a constant level while the other plate is forced upward over it.) Many of these thrust faults occur fairly near the surface. Because of their relative nearness to the surface and because the amount of energy released is very great, the earthquakes generated by this faulting are frequently *very* intense, creating severe surface

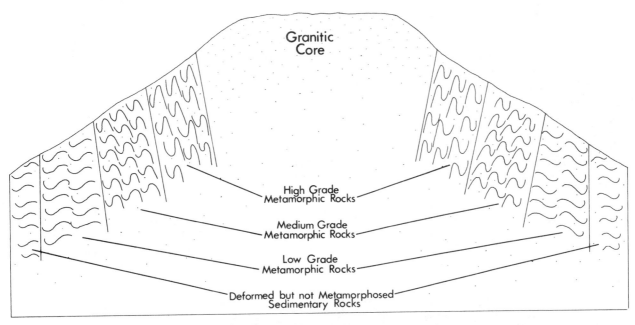

Figure 10.6. Diagrammatic view of the core of an eroded folded mountain belt.

disturbances and, if near populated areas, severe destruction and loss of life.

Because the intensity of deformation is so great, folded mountain belts are characterized by extensive metamorphism. Unlike the blueschist metamorphism of the ocean-ocean or ocean-continent collisions, metamorphism associated with continent-continent collisions is characterized by progressively increasing pressure and temperature. This type of metamorphism has been given many names, but let's call it "regional metamorphism." Obviously, the highest degree of regional metamorphism (highest temperatures and pressures) will be found deep within the continental masses and nearest the suture zones. The high pressures are generated by the collision and by the weight of overlying rock material. The high temperatures are found only at depth, mostly the result of deep burial of rocks. A relatively small contribution of frictional heat is generated by the collision process. Progressively higher (nearer the surface) or farther from the suture zone, the degree of metamorphism decreases until generally far from the suture zone, the rocks are still structurally deformed (folded or faulted) but show no evidence of metamorphism. Figure 10.6 shows the relationship between metamorphism and igneous activity in the eroded core of a folded mountain belt.

The actual cores or deepest, most intensely deformed parts of folded mountain belts are very frequently characterized by a great deal of *granitic* rock. Where does the granitic magma come from? Certainly not from any pressure-release mechanism and apparently not from any subduction process. However, these granites *do* seem to be the result of partial melting. In this case, the partial melting appears logically to be the result of very high-temperature and high-pressure metamorphism. As the conglomeration of rocks in this most intensely deformed and metamorphosed zone are heated beyond about 650°C, a granitic melt will form. This has been verified by laboratory studies. If we mix samples of all kinds of rock in continental crust, combined in appropriate proportions and subject them to increasing temperatures, the *partial* melt produced is, in fact, granite-like (granitic).

When the cores of folded mountain belts are exposed by weathering and erosion of overlying rock, we see frequently a granitic *igneous* core surrounded by high-temperature–high-pressure metamorphic rocks (intensely deformed), in turn surrounded by metamorphic rock, indicative of progressively lower temperatures and pressures. The actual suture zones are very difficult to delineate, however. One might think that the granitic cores delineate the suture, but even parts of the granitic cores get faulted and transported later in the deformational event, making the identification of suture zones a near impossibility.

CHAPTER 11
Hot Spots
Magma not necessarily associated
with plate boundaries

In the previous two chapters, we discussed magma formation at divergent and convergent boundaries. That discussion accounted for most of the magmatic activity in the earth's lithosphere—basalt at divergent boundaries, andesite-diorite in the volcanic-plutonic arcs associated with oceanic lithosphere convergence, and granite cores in folded mountain belts caused by continent-continent convergence. There exists, however one other category of magmatic activity—hot-spot magmatism, magmatic activity that can occur either within plates or on plate boundaries. This type of magmatic activity is responsible for such phenomena as the Hawaiian Islands, Iceland, and perhaps Yellowstone National Park, to mention a few.

Hawaiian Islands—Emperor Seamounts

Not all magmatic activity is restricted to plate boundaries, and not all is caused by plate interactions. For a long time after the "discovery" of plate tectonics, geologists puzzled over such features as the Hawaiian Islands, a site of active volcanic activity within the Pacific Plate (see Figure 11.1). It was not just the geographic location of the islands that was puzzling, it was also the composition of the lava. Unlike most of the volcanic islands in the Pacific that erupted *andesite,*

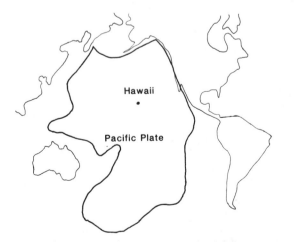

Figure 11.1. Location of Hawaiian Islands relative to the Pacific plate.

the Hawaiian Islands were composed of **basalt.** Apparently this extrusion of basalt is not really a recent occurrence, either; the islands show evidence of basaltic volcanism as long ago as 40 million years. Finally, again unlike other Pacific volcanic islands, the Hawaiian Islands are not associated with a deep-ocean trench.

The puzzlement continued until a geologist, Jason Morgan, of Princeton University, postulated a theory to account not only for the origin of the Hawaiian Islands, but also other centers of volcanic activity. Morgan postulated that within the asthenosphere there existed several stationary sites that almost continuously produced basaltic magma. He called these centers of volcanic activity **hot spots.** Morgan proposed that periodically basaltic magma would be produced, force its ways through the lithosphere and extrude into the surface to form broad, relatively low-lying lava mounds. (We do not mean to suggest that this is a small-scale event. Figure 11.2 is a cross-section across the big island of Hawaii. Note that although the volcanic island is not very steep-sided, it is quite high, greater than 28,000 feet, from the base of the ocean floor and is quite broad and therefore constitutes a very large volume of basalt.) Morgan further suggested that as the lithosphere moved across the asthenosphere, the hot spot would periodically extrude basalt onto the lithosphere, creating a chain of islands (see Figure 11.3).

This theory has been supported by radiometric dating of the lavas of other, dormant, Hawaiian Islands. The only active volcanoes are on Hawaii, the big island, southernmost of the chain. Radiometric dating of the lavas on the other islands show that the age of the islands progressively farther from Hawaii are progressively older. Figure 11.4 is a graph showing the age versus distance from Hawaii. If Morgan's hot-spot hypothesis is correct, it should be possible to determine from this type of graph the velocity at which the Pacific Plate is moving. These calculations suggest a velocity of about 10 cm/year, a velocity that is comparable to velocities determined by other methods.

A further validation of Morgan's hypothesis comes from the recognition that the northwesterly direction of movement implied by his hypothesis is also the di-

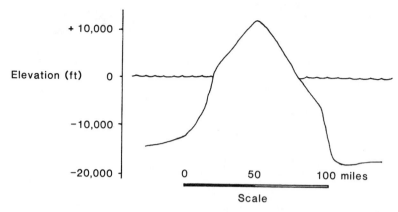

Figure 11.2. Profile across island of Hawaii.

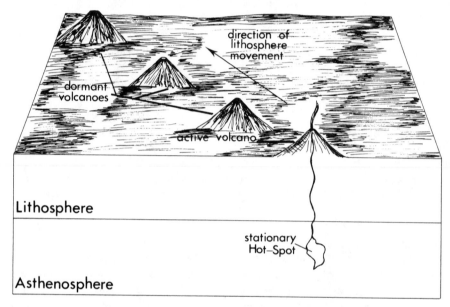

Figure 11.3. Origin of hot spot volcanoes. Lithospheric plate moves across a stationary hot spot in mantle. Lava erupts periodically producing a line of volcanoes.

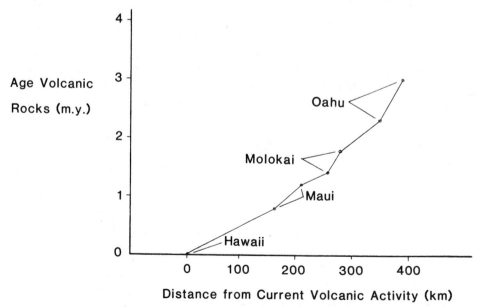

Figure 11.4. Relationship between age of lava on Hawaiian Islands and distance from island of Hawaii.

rection of movement of the Pacific Plate determined by other evidence.

Thus, the hot-spot hypothesis seems to account for the origin of the Hawaiian Islands. However, a few "little" problems remain. What is the process by which the magma is generated? It certainly must be by some process involving partial melting of the asthenosphere; after all, the magma is basalt. But basalt produced at spreading centers, or divergent boundaries, is apparently the result of pressure reduction in the asthenosphere due to cracking of the lithosphere at the spreading center. This is obviously not the process by which Hawaiian Island basalt is produced, because there is no apparent pressure-reduction process going on the middle of the Pacific Plate. That leaves only increased temperature as a process to produce a basaltic partial-melt magma. But these "centers" of volcanic activity are very small; in fact, they can almost be considered point sources of magma. Why would an isolated spot in the asthenosphere be hotter than the surrounding material, and even more important, how can it *remain* hotter over a period of millions of years? One would assume that the temperature at any given level in the asthenosphere would be nearly constant, and even if by some process one spot got hotter, the temperatures would soon equalize out. Obviously whatever is causing the "hot spot" to be hotter than the surroundingss must be an ongoing process. But the nature of the process still eludes geologists.

Referring back to Figure 11.1, you can see that another chain of mountains, the Emperor Seamount Chain, appear to butt up against the northernmost of the Hawaiian Islands. These seamounts are isolated mountains on the ocean floor whose tops are now below sea level. This particular seamount chain trends in a north-south direction. When studied closely, it was revealed that these seamounts were in fact also dormant basaltic volcanoes. The ages of these seamounts are progressively older from south to north. The southernmost of the Emperor Chain is about 40 million years, approximately the same age as the oldest of the Hawaiian Islands, and the northernmost is about 75 million years.

Iceland

Iceland, with its basaltic lava flows, sits atop the Mid-Atlantic Ridge. As such, its basalt could be considered to be produced by pressure-reduction partial melting of the asthenosphere below a spreading center. However, based on subtle but significant differences in the chemical compositions of Icelandic lavas from other spreading-center basalts, many geologists now believe that Iceland sits atop a mantle hot-spot that apparently only coincidentally lies directly below a spreading center. Thus the volume of basalt produced on Iceland may

be explained by the fact that two different processes, spreading center volcanism and hot-spot volcanism, are acting at the same place.

Other Hot Spots

It has been suggested that other centers of basaltic magmatic activity might also represent hot-spots. These have been studied to a considerably lesser degree, however, and not everyone agrees that each is caused by hot-spot volcanism. These other postulated hot-spot areas include the Galapagos Islands in the Pacific (coinciding with a spreading center), the Azores Islands in the Atlantic (coinciding with the junction of two plate boundaries), and the Yellowstone National Park area in northwestern Wyoming (an intraplate area).

Hot Spots as Initiators of Rifting

Finally, it has been suggested by some geologists that hot-spots, especially if they occur within plates and under continental lithosphere, are the first stage in the development of a new rifting zone that could lead ultimately to the breakup of the continental mass. Hot-spot activity, under these circumstances, is believed to cause the continental area to bulge upward, generating three cracks at approximately 120° angles to one another. As uplift and rifting continue, eventually the continental mass may crack along two of the cracks creating a new ocean basin. The splitting of the Saudi Arabian Plate from Africa is thought by some to be an example of this type of process (see Figure 11.5). The two cracks grew into the Red Sea and the Gulf of Aden. The third crack which is considered not to have developed further, is an extension of the East African Rift Valley.

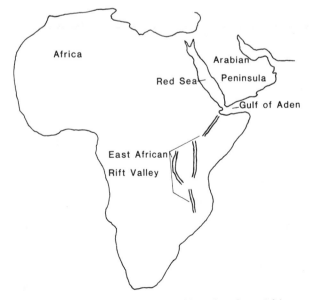

Figure 11.5. Rifting of Saudi Arabian plate from African plate.

67

CHAPTER 12
Weathering, Erosion, Transportation and Deposition

Those processes that are restricted to the surface of the earth—weathering, erosion, transportation, deposition and landform development—are generally not thought of as being so closely tied to plate motions as the internal processes discussed in chapters 7–10. Indeed, weathering, erosion, transportation and general characteristics of deposition could be looked at as isolated processes having no bearing on plate movements. The process of deposition or sedimentation, however, *can,* and, to our way of thinking, *should* be looked at in a plate tectonics framework. We will begin with a general discussion of surface processes and follow that with a chapter showing the plate tectonics influences on sedimentation.

Weathering

Weathering of a rock or a mineral is the processs by which the rock or mineral is brought into equilibrium with the pressures and temperatures at the surface of the earth. It involves the disintegration and/or decomposition of the rock or mineral. Weathering as a process is important because it yields as its products the raw material from which sedimentary rocks are formed. Weathering affects all rocks, even those sedimentary rocks formed at the surface. When discussing weathering, it is good to define two products of weathering: regolith and soil. **Regolith** is loose rock debris on the present surface. **Soil** is the uppermost part of the regolith that has been acted on by physical, chemical and biological agents; it is that part of the regolith that can support rooted plants.

Weathering is subclassified on the basis of what happens to the rock. Two subclasses of weathering are **physical weathering** and **chemical weathering.** We will examine the processes and products of both types of weathering.

Physical weathering involves the physical breakdown of rock to smaller pieces of rock. It is the process by which little ones are made from big ones. Several factors affect rock and aid in physical weathering:

1. **Jointing.** Pressure-release cracks in rocks; there is no relative movements of the two surfaces except movement away from one another.

2. **Thermal contraction** and **expansion.** Due to temperature changes; probably not very significant in a short term sense; perhaps more significant in longer term sense, especially if some fatigue (weakness) sets in over time.
3. **Frost action.** Due to the fact that water expands when it freezes, this can actually pry rocks apart when water enters a fracture and freezes.
4. **Work of plants or animals.** Roots seeking soil and water can pry rock apart. This category also includes the aerating work of worms in soil and rock.

Chemical weathering is the response of a rock or mineral to oxygen and water at the surface. Other chemical agents include carbon dioxide and acids. There are several processes by which chemical weathering takes place.

1. **Oxidation.** Reaction of rock or mineral with oxygen.
2. **Hydrolysis.** Reaction of rock or mineral with water.
3. **Dissolution.** Reaction by which all or part of the rock or mineral dissolves.

Climate is very important to chemical weathering. If, as stated above, chemical weathering is the response of a rock or mineral to oxygen and water at the surface, it is easy to understand why rocks decompose faster in areas of high rainfall. Temperature also plays an important role. Warmer temperatures promote faster decomposition. So, therefore, rocks decompose fastest in warm and humid areas.

In general, the products of chemical weathering fall into four categories, all four of which are chemically or physically stable at the earth's surface:

1. **Quartz.** A very stable (physically) mineral relatively resistant to chemical attack.
2. **Clays.** Very stable chemically, the products of the decomposition of aluminum-bearing parent materials; very fine-grained and chemically stable.
3. **Solutes.** Those materials which go into solution; generally including sodium (Na), potassium (K) and calcium (Ca); some silica (SiO_2) goes into solution, as does small amounts of almost all chemical constituents.

4. **Iron oxides and hydroxides.** The products of the chemical decomposition of iron-bearing minerals like pyroxene and amphiboles; a reddish-brown or reddish-yellow powdery substance that causes the typical red or yellow color in chemically weathered rock.

Table 12.1 shows the chemical weathering products of five fairly common rocks: granite, basalt, quartz sandstone, limestone and shale.

In Table 12.1, we give chemical **end-products** of the weathering of various rock types. Frequently, however, chemical weathering does not proceed to completion, and some amount of material that is not completely chemically altered is removed, finds its way into the transporting medium and even to the site of deposition. The proportion of chemically stable to chemically unstable mineral or rock components (especially feldspar and rock fragments) defines the **maturity** of sediment. A very mature sediment is one with very little unstable material. An immature sediment is one with a relatively high proportion of unstable material. The maturity is determined by a complex interplay of physical and chemical weathering and erosion. In general, a higher degree of physical weathering promotes a higher degree of chemical weathering. Physical weathering produces more surface area by breaking large rock

---

Table 12.1. Chemical Weathering Products of Various Rock Types.

GRANITE

Original principal minerals include two feldspars, quartz and a dark iron/magnesium mineral.

Potassium feldspar breaks down to clay, with the potassium going into solution.

Plagioclase feldspar also breaks down to clay with the sodium and calcium going into solution.

Quartz remains as quartz, although a small amount of silica may go into solution.

Iron/magnesium silicates break down to iron oxides and hydroxides; aluminum (if present), magnesium, and calcium (if present) go into solution; and aluminum and silica form clay.

Mica (if present) will ultimately degrade to clay.

The end-product of the chemical weathering of a granite in a warm, humid climate is a sandy, red-yellow clay-rich soil.

BASALT

Original principal minerals include plagioclase feldspar, pyroxene and olivine.

Plagioclase feldspar breaks down to clay with the calcium and sodium going into solution.

Pyroxene and **olivine** both will break down to iron oxides and hydroxides; calcium (if present) and magnesium go into solution; the aluminum and silica form clay.

The end-product of the chemical weathering of basalt in a warm, humid climate is a dark red clay-rich soil with little or no sand.

QUARTZ SANDSTONE

Original principal mineral is quartz cemented by other quartz (silica).

Quartz remains as quartz; some of the cementing material goes into solution, loosening the quartz grains.

The end-product of chemical weathering of quartz sandstone in a warm, humid climate is granular quartz, with no clay. If there was even a small amount of iron/magnesium minerals in the quartz sandstone, the granular quartz will be stained red by iron oxides and hydroxides.

CLAY-RICH LIMESTONE

Original principal constituents include calcite, clay and some organic matter.

Calcite goes into solution if the water is even slightly acidic, yielding calcium in solution and carbon dioxide to the atmosphere.

Clay remains as clay.

Organic matter may decompose somewhat.

The end-product of chemical weathering of a clay-rich limestone is an organic and clay-rich soil.

SHALE

Original principal minerals include clay and fine-grained quartz. Some shales also contain calcite and/or organic matter.

Clay remains as clay.

Quartz remains as quartz.

The end-product of chemical weathering of shale in a warm, humid climate is a clay-rich soil with some fine-grained quartz. This soil will not differ significantly from its parent shale, except to be disaggregated.

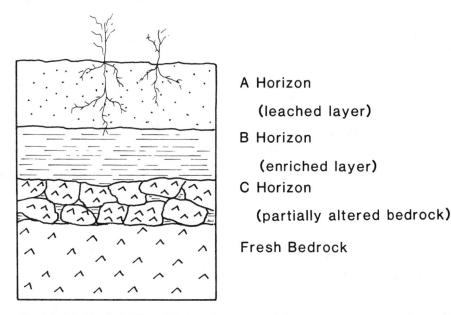

A Horizon

    (leached layer)

B Horizon

    (enriched layer)

C Horizon

    (partially altered bedrock)

Fresh Bedrock

Figure 12.1. Typical soil profile showing A, B and C horizons relative to fresh bedrock.

masses into smaller rock masses. More surface area gives the agents of chemical weathering (especially water and oxygen) more area on which to work, thus speeding up the chemical weathering process.

One must also consider the land elevation of the rock being weathered relative to sea level. The higher the **relief** (difference in elevation from highest to lowest points), the faster material gets stripped away and, therefore, the more immature the sediment. Conversely, if the source of the weathered material is at a relatively low elevation, it stays longer in the source area where it is subjected to chemical weathering and will be more likely to produce a chemically mature sediment. The maturity of a sedimentary rock is an important indicator of the conditions that existed when that sedimentary rock was formed. By examining the maturity of an old sedimentary rock, we can deduce something about the tectonic conditions which must have been present in the past, when direct evidence for the conditions no longer exist.

Finally, in this discussion of weathering, let's examine the formation of soil. True soil is a surface or near-surface component. The thickness of soil and its degree of development depend upon many factors—climate and rock-type especially. If we examine a typical soil profile, as in Figure 12.1, we see several "zones" of soil depending on the degree of alteration of the parent rock. Nearest the surface, the **A horizon,** the soil has been leached of soluble material. Humus (organic matter) from decayed plants has been incorporated. The leached material has migrated downward, and some clay has also been transported downward. The next zone is the **B horizon** which is clay enriched. This is where the material from the A-horizon accumulates; it is generally more dense than A-horizon and more en-

riched in clay. The A and B horizons constitute true soil. The **C horizon** is weathered bedrock. It is not good soil; rather it is the transition zone from rock to soil.

In earlier times, it was thought that the composition of bedrock was the major factor in determining the type of soil that developed. In more recent times, it is thought that the *climate* in which the soil develops is the most important factor. Therefore, different rocks in the same climatic zone tend to yield similar soils; whereas, the same rock weathering in two different climatic zones tends to yield two different soils.

Erosion and Transportation

Erosion is the process by which weathered material at the earth's surface is stripped away from where it is formed and put into movement to be transported to its ultimate depositional site. The principal agents of erosion and transportation are **water, ice** and **wind.** Erosion is also a sculpting force which produces many land forms on the earth's surface. In this chapter, we will examine the processes of erosion and transportation as they affect the movement of sediment from its source area to its site of deposition. In Chapter 13, we will see that the type of sediment that accumulates in a given plate tectonic environment has unique properties. When these properties are seen in an ancient sedimentary rock, we can deduce something about the sedimentary environment in which it formed and the tectonic conditions at the time of its formation.

Action of Water

Streams and rivers probably erode and transport more material than the other two media together. Running water has the ability to erode material over which or past which it flows. In general, the faster the water is moving, the more it is able to erode and transport. Running water carrying sediment is an even more destructive force because the particles carried by the stream can abrade the sides and bottom of the river or stream channel. A stream will continue to erode until it is carrying its maximum sediment load.

When considering the sediment carried by a stream or river, we need to consider three components of its load: the **dissolved load,** the **suspended load,** and the **bed load.**

The **dissolved load** is that part of the total load that is in true solution. It consists of chemical ions produced by chemical weathering. Because dissolution of most minerals is a slow process, streams and rivers rarely carry their maximum dissolved load. World-wide, the amount of stream load carried as dissolved load is about 20% but can reach as high as 50%. The dissolved load plays no part in the erosion that a stream can cause.

The **suspended load** is that part of the total load that is carried in suspension. Even though most sediment is more dense than water (and therefore would settle to the bottom in still water), most streams are quite turbulent, and this prevents the sediment from settling. Stream turbulence is influenced by the velocity of the stream and the nature of the stream channel. Faster flowing water is generally more turbulent water. A rough or uneven channel also increases the turbulence. As turbulence increases, the amount of sediment in suspension increases. Most rivers carry most of their total load in suspension.

Bed load is that part of a stream's load that is carried along the stream bottom by rolling, sliding or bouncing. It contains the largest particles that are being moved by the stream. The maximum particle size able to be carried by a stream is known as its **competence** and is a function of stream velocity. However, particle shape and density also influences its ability to be transported. Rounded particles are easier to transport than angular particles. Less dense particles are easier to transport than more dense particles.

To put this all together, visualize an ideal river system that begins as a mountain stream and ends as a large river entering the ocean. At the headwaters of the river, in the mountains, the ground slope is quite steep, the river velocity is quite high, and the water is quite turbulent. The total amount of sediment carried by the mountain stream is relatively small, but it has the ability to transport very large boulders. As the ground slope decreases and width of the river channel increases, the velocity of the stream also decreases as does its ability to transport the largest boulders. The water is less turbulent and slower moving, but because the volume of water has increased from contributions of tributaries along the way, the total suspended load is vastly increased. The bulk of the load being transported by an ideal stream is therefore found near the mouth.

Action of Ice

The second agent of erosion and transportation we will discuss is ice in the form of **glaciers.** Glaciers exist in two forms:

1. **Continental ice sheets.** Very broad and generally restricted to high latitudes (nearer the poles); they are more extensive, of course, during an Ice Age. Present examples include Greenland and Antarctica.
2. **Mountain (or valley) glaciers.** Restricted to existing mountain valleys at high altitude and medium to high latitudes.

A glacier is the most powerful erosive agent and has the ability to carry the largest sized material. Volume for volume, glaciers can also carry the largest amount of sediment. Both kinds of glaciers erode by **plucking** (pulling off loosened blocks from valley walls and glacial floors) and by **abrasion** (scraping of underlying rock by sediment imbedded in the ice).

Action of Wind

The third agent of erosion and transportation, and generally the least effective, is **wind.** Wind can carry only relatively fine material and, in general, not a large volume of material. Both the size of the particles that can be carried and the amount of material that can be carried are a function of wind speed. Wind usually picks up previously loosened material rather than actively loosening and removing. Obviously, to anyone ever caught in a windstorm, the erosive power of wind can be very strong if the wind is already carrying sediment. Sand blasting is a very effective natural, as well as artificial, erosive agent.

Wind moves particles in two ways:

1. **Suspension.** As in the case of wind storms.
2. **Saltation.** Analogous to stream bed load where a sand grain is pushed along the surface by wind and is temporarily picked up into the air; if when the grain falls back to the surface, it strikes another grain, the second grain may be kicked into the air temporarily.

In general, smaller sized material is carried in suspension, and larger sized material by saltation, but both particle size and amount of material carried increase with wind velocity and turbulence.

Roundness and Sorting

Before leaving this section on erosion and transportation, let us consider two sediment parameters that are primarily affected by erosion and transportation: **grain roundness** and **sorting.**

Weathering and erosion of source rocks yields grains that frequently are quite angular in shape. To a large extent, the angularity of grains depends on the nature of the source rock and the particular mineral being released; igneous and metamorphic source rocks, in general, yield more angular grains than many sedimentary rocks. In the process of erosion and especially transportation, the angularity of sediment particles is decreased if the sediment has the opportunity to collide with other sediment particles. In general, sediment that is transported by water decreases in angularity (and hence becomes more round) with length of time in transit, i.e., the farther a sediment particle must travel, the more rounded it is likely to become.

Most information concerning roundness comes from sediment transported by water, primarily because water moves such a large volume of sediment. One can say that because glacially transported sediment is surrounded by ice, and therefore protected from abrasion by other particles, there is no significant increase in roundness in glacially transported sediment, except possibly for that carried near the floor of the glacier. Sediment transported by wind will undergo little abrasion, but it may abrade the sediments it is blown over. However, due to the small size of most material carried by wind, little abrasion or rounding will take place. Rather, both abraded and abrading grains in a wind environment may take on a surface etching or frosting.

Sorting refers to the range in grain sizes in a body of sediment or sedimentary rock and can tell you something about the transportational and depositional history. Ice keeps everything from large boulders to finely ground rock flour in suspension and, therefore, sediment that has been transported by ice can be very poorly sorted. Because water can keep only a limited range of sizes in suspension, depending on the stream velocity and turbulence, sediment that has been transported by water tends to be better sorted. Finally, wind can keep only a very small range of grain sizes in suspension, depending again upon the velocity and turbulence of the wind. Therefore, material transported by wind tends to be better sorted than material transported by either ice or water.

Deposition

Deposition occurs when the transporting medium can no longer transport all or part of its load. In this section, we will look only at the general aspects of deposition, but we will return to examine in much more detail specific depositional environments in Chapter 13. Again, because the vast majority of sediment is transported by water, we will look mostly at what conditions cause running water to deposit its load, with less emphasis on deposition of sediment transported by ice and wind.

We have already alluded to the factors that govern deposition of sediment transported by water. Deposition from running water is a function of changes in the stream velocities and stream turbulence. In mountainous areas, where stream velocities are high, and the water very turbulent, large particles are moved in the stream's bed load. As the slope of the land flattens, stream velocity and turbulence decrease, and the stream is no longer capable of moving the largest material, and it is left behind, generally in the stream bed. As the slope of the land continues to flatten, progressively smaller particles are left behind. However, most deposition of suspended material takes place when the river or stream enters a standing body of water, such as the ocean or a lake. The velocity of the stream decreases rapidly as it enters the still-standing body of water. In general, the coarsest part of the suspended load drops out first, nearest the shore, and progressively finer material is deposited farther from the shoreline. Using these principles and assuming no reworking of the sediment after deposition, it is possible to determine ancient shorelines in the rock record by examining the average size of various sedimentary rock particles deposited simultaneously. It must be remembered, however, that material removed from suspension can be altered and even transported in the depositional environment, and the possibility of later alterations must always be kept in mind in trying to deduce ancient sedimentary environments.

CHAPTER 13
Sedimentary Environments

In this and the next chapter, we will discuss the processes and products of sedimentation in a plate tectonics framework. This is a different approach from that usually taken in introductory geology texts that tend to treat this process as a surface process (which it is) with no relationship to the tectonic setting. We believe that the "big picture" of sedimentation can best be made clear to introductory students by putting sedimentation into a plate tectonics setting. This is the approach being taken by modern sedimentologists, but it has not found its way into the introductory courses—except this one.

The sediment that is deposited in any given depositional environment is a result of a complex interplay of both physical and chemical processes. To a large extent, the physical processes involved in sedimentation govern the physical characteristics of the sediment itself—average grain size, sorting, vertical and horizontal variation in grain characteristics, with essentially no regard to the mineralogical makeup of the sediment. In other words, it makes no difference what minerals constitute the sediment being transported and deposited, only the physical properties of grain size, grain shape and grain density seem to have a bearing on how and where the sediment is deposited. Further, to some extent, the *physical* characteristics of a sediment in a given depositional environment are not even a function of the specific tectonic setting adjacent to the depositional environment.

The *mineralogical* composition of the sediment is, however, related directly to the tectonic setting of the source of the sediment. Specifically, the minerals themselves reflect the source area, and the maturity of the sediment reflects its tectonic setting.

As a result, we can see sand-sized sediment derived from sources in several tectonic settings. The mineralogical composition and maturity of these sand-sized bodies of sediment will differ depending on the tectonic setting of the source area. Likewise, we can find shales being deposited in several depositional environments, differing only in mineralogical composition and maturity. The same holds true for conglomerates and most other clastic sediment deposits.

In this chapter, we will discuss various depositional environments and the physical characteristics one might expect of the sediment being deposited in each. Because, however, it is possible to find similar depositional environments related to vastly different source-area tectonic settings, we will then, in the next chapter, examine various tectonic settings to see what depositional environments can exist, the relative importance of each and the compositional characteristics of each.

The goal of the geologist is to be able to interpret the depositional environment *and* tectonic setting in which an ancient sedimentary rock was deposited. Throughout the following discussion, we will use the principle of Uniformitarianism, "The present is the key to the past," to understand what governs the physical and chemical characteristics of modern sediment accumulation. We make the assumption that the same processes were acting in the past, and that a sedimentary rock with physical and chemical characteristics similar to a body of sediment now being deposited was probably a result of these same processes (physical and chemical) at some time in the past.

Sedimentologists have defined several depositional environments. In this chapter, we will examine only those that comprise a major portion of the sedimentary record:

1. Shallow marine
2. Continental slope
3. Turbidite
4. Abyssal plain
5. Non-marine

Shallow Marine Depositional Environments

Offshore Clastics

By shallow marine, we mean an environment extending from the coastline to a depth of about 200 meters. In this category, we find **continental shelves** (those extensions of the continents presently under water, see Figure 13.1) and **epicontinental (epeiric) seas** (those incursions of ocean onto the continental interiors in times of elevated sea levels.) Recall from Chapter 12 that sea-level is an ever-changing plane, depending primarily on how much sea water is tied up in continental glaciers with lower sea-level during Ice Ages, higher sea-level during **interglacial periods.**

Because sea-level is constantly changing *and* because the land surface covered by and adjacent to shallow marine seas is generally of low relief, small changes in sea level can cause large differences in where shore-

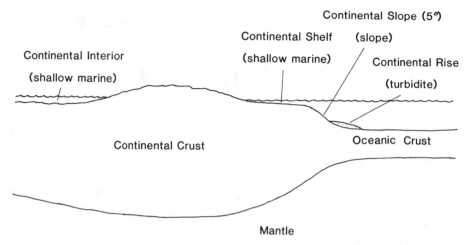

Figure 13.1. Typical relationship between continental shelf (bordering deep ocean) and continental interior. Sedimentary depositional environments are noted in parentheses.

lines are located. This, in turn, can have a profound effect on the distribution, both laterally and vertically, of the sediment being deposited, because characteristics of clastic sediments predictably vary with distance from the shoreline.

One of the key influences on the sedimentation of clastics is the climate and relief of the source area supplying the sediment. If the source area is of high relief, we must also consider the proximity of the source area to the shoreline. Consider first a high-relief source area near the shoreline. This situation will generally supply large volumes of sediment to the depositional environment. (Climate in the source area also is important; high rainfall generally increases the amount of sediment supplied.) Most of the clastic sediment will be transported by rivers, and deposition of the stream load will follow the general rules established earlier: The faster and more turbulent the stream, the coarser its load can be. Reduction of either stream velocity or turbulence will lead to the gradual deposition of progressively finer-grained material from its suspended load. Close proximity of a high-relief source area to the shoreline can bring about the deposition of very coarse-grained clastic sediments at the shoreline.

If, however, the source area is of low relief (or if the high-relief source area is very far from the shoreline), a lower volume of sediment will be supplied to the depositional environment, and the grain size of the coarsest sediment supplied would be smaller. In these cases, most of the sediment will be sand-, silt- and clay-sized, resulting in the deposition of sandstones, siltstones and shales.

As the clastic sediment is delivered to the sea, much of the material will be deposited at or very near the shoreline, a process that should yield a poorly sorted clastic rock. Examination of present-day shoreline deposits and ancient sedimentary rocks thought to be deposited under similar conditions, however, show a pattern of laterally extensive (extending far out from the shoreline), well-sorted and thin-layered clastic sedimentary rock. Apparently, even though poorly sorted material is deposited near a shoreline, this material gets reworked by wave action after deposition. The constant barrage of waves **winnows** (separates out) the finer-grained material, leaving well-sorted, clean sand-sized material at the shoreline. The silts and clays which are winnowed out from the sand are transported away from the shoreline by the waves. The clays are transported the longest distance and are deposited farthest from the shoreline. The silts, being coarser grained than the clays, are not transported as far and are deposited at an intermediate distance from the shoreline. If the relief of the area on which the sediment is being deposited is low, this process will yield a sediment package that is laterally extensive, but with an average grain size that varies laterally, coarsest near the shoreline and progressively finer away from the shoreline. Figure 13.2 shows how such a sediment package would look diagrammatically.

In the rock record, geologists see laterally continuous beds of sandstones, siltstones and shales, much more extensive (i.e., covering wider areas) than expected based on the above model of shallow marine clastic sedimentation. Until recently, a laterally continuous bed of sandstone (or siltstone or shale) was interpreted as being **time equivalent,** i.e., all of the sandstone was interpreted as having been deposited within the same period of time under what must have been nearly identical conditions (same source material, same transportational history, same shoreline proximity, etc.) This created a problem for many geologists because there were no present depositional environments where these very extensive bodies of sandstone (or siltstone or shale) were being formed. In other words, we had no present model on which to base our model of the past. We now realize that these **lithologically continuous** (same rock composition and environmental implications) bodies might *not* have been deposited simultaneously, even

76

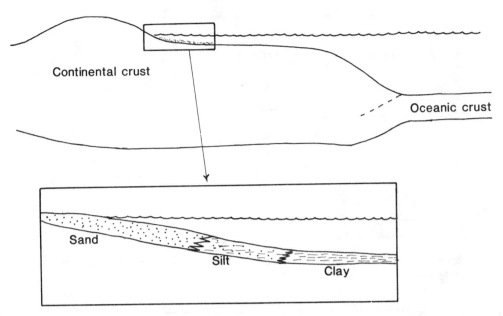

Figure 13.2. Diagrammatic sketch of offshore clastic sediment deposited in shallow marine environment. Note decrease in grain size away from shoreline.

though they are laterally continuous. Consider Figure 13.3 which shows the same sequence of sedimentary rocks as in Figure 13.2. The dashed line represents a higher sea level than in Figure 13.2. With the sea level in this new position, and assuming the same type of sediment being delivered to the shoreline, new sedimentary units would be deposited in slightly different locations relative to the units deposited at the lower sea level: New sandstones would be deposited further "inland," and new siltstones would form, at least in part, directly above where sandstones were previously deposited. New shales would be deposited, at least in part, directly over previously deposited siltstones. Thus, if sea level were rising slowly and continuously, a pattern of sedimentary rocks would be deposited over time as shown in Figure 13.4. In this figure, it can be seen that—at any given time—sandstone, siltstone, and shale were being deposited simultaneously. Over a pe-

riod of time, as sea level rose, the relative positions where sandstones, siltstones and shales were deposited shifted, giving rise to what appears to be continuous, laterally extensive beds of sandstone, siltstones and shales. In fact, however, a given layer of sandstone (or siltstone or shale) represents deposition at different times, in different locations, but under almost identical physical conditions.

The sequence shown in Figure 13.4 is known as a *transgressive* sequence. A similar pattern, oriented differently, would be seen if sea level were *dropping* slowly and continuously, forming a *regressive* sequence, shown diagrammatically in Figure 13.5. Thus, it is possible to deposit laterally continuous thin sandstones, for example, over a period of time, reflecting a slowly migrating shoreline due to slow and continuous changes in sea level.

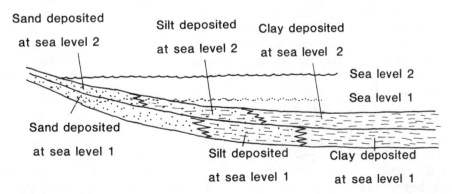

Figure 13.3. Diagrammatic sketch of a later offshore clastic sediment packet in a shallow marine environment deposited at a higher sea level.

77

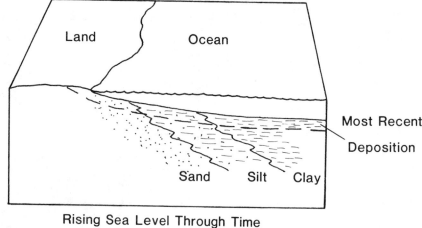

Rising Sea Level Through Time

Figure 13.4. Block diagram showing the development of a transgressive clastic sequence deposited during period of rising sea level.

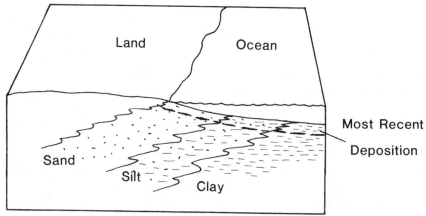

Falling Sea Level Through Time

Figure 13.5. Block diagram showing the development of a regressive clastic sequence deposited during period of falling sea level.

Nearshore Environments: A Slightly Different Story

While the above discussion of deposition in a shallow marine environment is accurate on a large scale, it oversimplifies the sedimentary processes at the shoreline. While it is possible to deposit mostly sand-sized sediment at the shoreline, the actual depositional processes active on shorelines is considerably more complex. Let us examine a couple of common subenvironments of a coastal shoreline: deltas and barrier-island/marsh environments. A delta is a triangle-shaped wedge of sediment deposited at the mouth of large, sediment-laden rivers (like the Mississippi and Nile Rivers, for example). When a river enters the ocean, most of the river's energy is dissipated as its progress is stalled. When a stream loses its energy, it can no longer keep its sediment load in suspension. This suspended matter drops out of suspension and is deposited. In general, the coarser-grained material is deposited nearer the shoreline, with progressively finer material being deposited farther from the shoreline.

Generally, deltaic deposits are "dumped" near the shoreline, building up a packet of sediment that rather abruptly terminates away from shore, producing a blunted front edge that is at a moderately steep angle to the bottom surface. A single packet of deltaic sediment is shown diagrammatically in Figure 13.6. As more sediment is supplied to the delta, sediment layers will build up on the top surface (producing **top-set beds**), out over the abrupt terminus of the previous deposits (producing **fore-set beds**), with some very fine-grained sediment being deposited over the previous bottom surface (producing **bottom-set beds**). These are shown diagrammatically in Figure 13.7.

As the top-set beds build, it is possible for them to grow up above sea-level, producing more "land" for the river to flow over. This new land surface will be very flat, and the river will have the tendency to meander over the top, taking one path for a while, and then migrating to where it takes an entirely different (see Figure 13.8) path. A river that migrates over these top-set

78

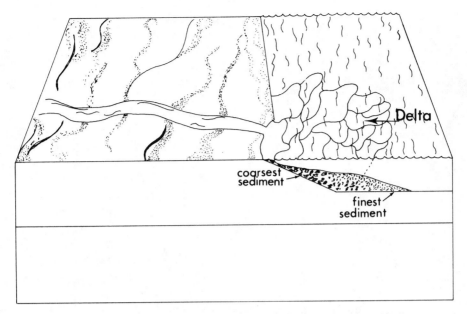

Figure 13.6. Diagrammatic sketch of a single packet of deltaic sediment.

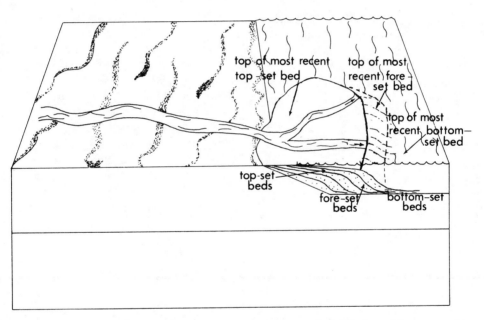

Figure 13.7. Development of top-set, fore-set and bottom-set deltaic beds as delta grows with time.

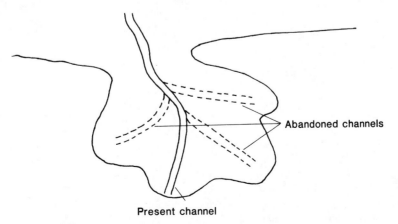

Figure 13.8. Bird's foot delta, showing abandoned channels.

79

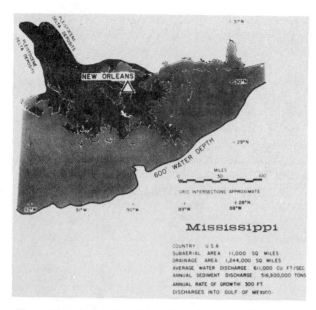

Figure 13.9. Mississippi River delta. (Photograph courtesy of Houston Geological Society.)

the river migrates to occupy a different channel. Old channel deposits can be overlain by new swamp deposits and *vice versa,* producing a complex packet of sediment that varies laterally, at any given time, and vertically over a period of time.

Frequently, seaward of the shoreline, we can see **barrier islands,** sand bars separated from the mainland by lagoons, as seen in Figure 13.10. Figure 13.10 also shows a **salt marsh subenvironment** behind the barrier island but part of the same environment. The origin of both barrier islands and marshes are controversial and beyond the scope of this book. We will only discuss the type of sediment deposited in this complex nearshore depositional subenvironment.

The barrier islands are almost always deposits of sand. They are generally well sorted because they are constantly being reworked by wind, which piles the sand up, forming **dunes.** These dunes are temporary features; they are constantly shifting. Some dunes become colonized by plants, which stabilize them and prevent their migration. Dunes can also be **breached** or cut through during storms, creating a tidal inlet (see Fig. 13.10). Behind the tidal inlet, it is not uncommon to find small tidal deltas, sand deposits pushed through the tidal inlet at high tides and deposited on top of lagoon sediments.

Behind the dunes, but still seaward of the lagoon, salt marshes are frequently found. Salt marshes are low-lying areas vegetated by salt-tolerant plants. They are periodically inundated by sea water at high tide and then exposed to the atmosphere at low tide. Marsh sediments are typically composed of clay-sized material and dead plant material. Even though they are exposed to the atmosphere at low tide, only the very top of the

beds will produce what's known as a **bird's foot delta.** The Mississippi River delta (Figure 13.9) is an excellent example of a birds-foot delta. When the river is flowing within its channel, most of its sediment will be deposited where the river enters the ocean. (A small amount may be deposited in the channel as sand bars.) When the river floods, however, and spills over its channel, even the silt and clay in the spill-over water will be deposited in lands and bay areas adjacent to the channel. Thus, we can expect to see a complex lateral sediment package being formed at any time in deltaic regions such as is shown diagrammatically in Figure 13.8. The picture becomes even more complicated when

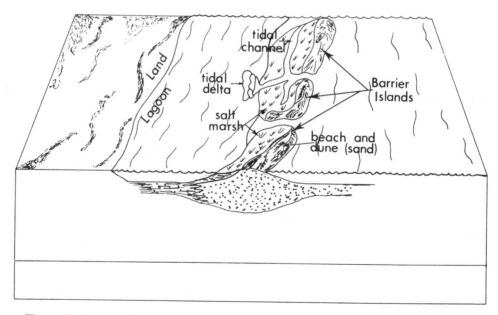

Figure 13.10. Block diagram showing relationship between barrier islands, salt marshes and lagoons in a nearshore environment.

marsh sediment is enriched in oxygen. Below about 1–2 cm, there is no free oxygen, and therefore organic material is not oxidized. Rather, it is degraded by bacteria, which produce H_2S ("rotten-egg") gas as a by-product.

The lagoons are generally areas of low wave energy, being protected by the barrier islands. Lagoonal sediments are generally finer-grained than the barrier sands, but perhaps coarser than the marsh sediments. Lagoonal sediments typically are made up of clay- and silt-sized material with some organic matter incorporated into it.

Thus, the barrier island/salt marsh/lagoonal subenvironment is another which is highly variable laterally. At any given time, sands (in the form of barrier-island sands), organic-rich clays (in the form of marsh sediments) and silty clays (in the form of lagoonal sediment) can be deposited simultaneously. Obviously, if sea level is either rising or falling, each of these features migrates either landward, if sea level is rising, or seaward, if sea level is falling. This creates an even more complex picture of both lateral (at a given time) and vertical (through a period of time) variability.

Carbonate Environments: The "Critters" Take Over

Consider next the effects of a *lack* of clastic sediment being delivered to the shoreline. One might simply expect no record of sedimentation during that period of time. If, however, the climate is warm, it is possible that **limestone** deposits will form. These limestones will almost certainly be of biologic origin and will take the form of limestone muds, limestone sands and/or organic reefs. Limestone sands are comprised of broken pieces of pre-existing shell material. There is one form of limestone sand that is apparently of inorganic origin. **Oolites** are spherical grains of calcium carbonate that grew by inorganic precipitation of concentric layers around some other "seed" particle.

Limestone muds were once thought to be inorganic in origin. It is now believed, however, that almost all of these fine-grained muds are biologic in origin. Most are thought to be the products of two processes: (1) decay of organic matter that holds very fine carbonate particles together in a shell and subsequent release of these particles and (2) secretions by algae. These lime muds can also be laterally extensive and thin-bedded. They are often interlayered with carbonate sands, but do not have to be so associated.

The final form of limestone we will discuss is the **organic reef.** Often, these are called coral reefs, but most are composed of several different types of colonial organisms, not only corals. Organic reefs, obviously are *not* thin-bedded and laterally extensive, but rather massive bodies of animal skeletal material that may extend for hundreds of miles (cf., the Great Barrier Reef of Australia). Figure 13.11 shows an example of a typical limestone depositional environment.

Evaporites: What Happens When the Water Gets Too Salty

Finally, continental interiors are often the site of deposition of the chemically precipitated sedimentary rocks known as **evaporites.** Evaporites take their name from the process by which they form. These rocks are restricted in the type of environment in which they form. Specifically, evaporites only can form when the dissolved salts in the sea water become so concentrated that the water is **chemically saturated,** i.e., the water

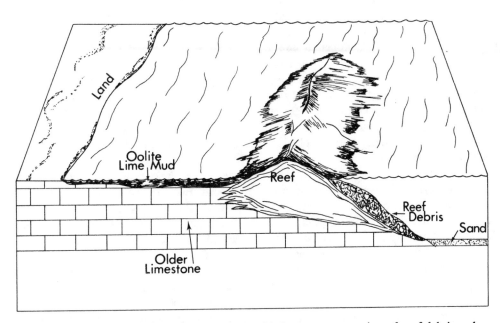

Figure 13.11. Block diagram showing relationship between an organic reef, reef debris and carbonate mud in a typical carbonate depositional environment.

81

can hold no more of a given salt in solution. However, because most sea water is very much *under*saturated, some process must occur to increase the **salinity** (total dissolved salts in sea water). Generally, an increase in the salinity takes place in a closed or partially closed basin where water is essentially trapped. Just as important, however, the rate of evaporation in the (partially) closed basin must exceed the input of water to the basin, so that a net decrease exists in the volume of water in the basin. Geologically, this might occur when epicontinental seas rise, filling a partially closed basin, then fall, trapping the water in the basin.

Evaporite sediments generally include halite (rock salt, NaCl), gypsum ($CaSO_4 \cdot 2H_2O$) and calcite ($CaCO_3$). Individual layers within an evaporite sequence are often **monomineralic,** i.e., all halite or all gypsum. The monomineralic nature of evaporite layers is explained by the fact that a given volume of water can dissolve different amounts of different constituents. Then, when sea water containing several dissolved constituents becomes increasingly more saline, saturation of particular constituents will take place at different salinities. When the water becomes saturated with one constituent, it is still undersaturated with respect to the other dissolved constituents. Each constituent will then precipitate from sea water at a given salinity, producing an essentially monomineralic rock. As the salinity continues to increase, a second dissolved constituent will reach saturation and begin to precipitate and a third and so forth.

Continental Slope Depositional Environments

The main feature governing deposition in continental slope depositional environments is that the deposition is taking place on an *inclined* surface. Most of the deposition we discussed in the previous section took place on horizontal (level) surfaces. **Continental slopes** are those inclined areas seaward of continental shelves. On the trailing edges of continents, the continental slope separates the **continental shelf** and the **continental rise** (the dropoff to the abyssal plain). On leading edges of continents, continental slopes separate continental shelves from the trenches. For many years, geologists believed that continental slopes were areas characterized by nondeposition and that any sediment found on the slope was in transport, on its way from the continental shelf to the continental rise. Now we believe that some deposition takes place on the slopes as well.

Most of the sediment deposited on the slope consists of silt- and clay-sized material of continental origin. This is the material that is carried out and over the continental shelf and is dumped onto the slopes. The sorting and amount of sediment deposited depends on several factors, but certainly includes the width of the shelf over which the sediment has travelled. The wider the shelf, the more material is deposited on the shelf,

and the less material is deposited on the slope. The sediment is typically deposited as inclined layers which frequently show signs of slumping and sliding.

Turbidite Depositional Environments

The term **turbidite** is given to a sedimentary rock which has been transported by a dense current of water and sediment known as a **turbidity current** and deposited principally on continental rises and in trenches. In reality, these sedimentary units can be deposited anywhere there is a relatively sharp break in the slope over which a turbidity current flows. One other distinguishing feature is that turbidites are the products of sporadic, or periodic, sedimentation. Deposition in shallow marine environments, on slopes and even in deep sea environments, is continuous. Turbidites are characterized by intermittent deposition followed by periods of nondeposition.

Apparently, the origin of turbidity currents involves an underwater buildup of sediment at the edge of a continental shelf. Periodically, this sediment pile is disturbed by earthquakes or slope failure or any number of other processes and begins to flow as a dense current of water and sediment, down the continental slope. Laboratory experiments involving turbidity currents and a few direct observations of turbidity currents demonstrate that these currents can flow as fast as several kilometers per hour, and they can maintain their physical integrity over very long distances. Being slurries of sediment and water, they can erode the slope over which they travel, adding additional material to the slurry. As with most other types of sedimentary environments, deposition takes place when the current begins to lose energy. This takes place principally when the slope on which it is travelling decreases, such as where continental slope meets the abyssal plain, or at the base of a trench.

The sediment that is deposited will be poorly sorted, composed of sediment of many size ranges, from pebbles to clay. The sediment will tend to be deposited in thick packets right at the slope break, i.e., these units will not be laterally continuous over larger distances. Because many turbidity currents flow down submarine canyons, they are unable to spread out until they reach the base of the canyon. This will result in a fan-shaped deposit. Because the decrease in slope where the continental slope meets the abyssal plain is usually fairly dramatic, the coarsest fraction of the sediment in the turbidity current will be deposited very quickly over a limited area. As this coarsest fraction is deposited, the progressively finer fractions are deposited both farther away from the base of the slope and on top of the previously deposited coarse material. A typical sediment packet of this environment is shown in Figure 13.12. Each turbidity current, then, deposits a sedimentary package that is coarsest nearest the base of the slope

Figure 13.12. Diagrammatic sketch of two turbidite sediment "packets." Note decrease in grain size laterally and horizontally.

and progressively finer-grained both vertically (over the coarsest fraction) and laterally (away from the base). An ideal sequence of sediment is said to show **graded bedding** (coarsest on bottom, progressively finer above). As mentioned earlier, these deposits are known as **turbidites.**

Each turbidity current has a limited period of time in which it is active. Following the initiation of a turbidity current and the deposition of its sediment load, is a period of nondeposition while another pile of sediment accumulates at the head of the submarine canyon. When something shakes this next sediment pile loose, another turbidity current is initiated, and it deposits its sediment load as a separate graded packet from the previous turbidity current. Thus, over a long period of time, a vertical sequence of several graded packets can be seen and is shown diagrammatically in Figure 13.12.

Abyssal Plain Depositional Environments

Abyssal plain deposits occur on the deepest part of the ocean floor from the continental rise outward away from continental sources. In general, these deposits consist of very fine-grained material that is deposited in thin, laterally extensive layers at relatively slow rates of sedimentation. Some of the material is the finest-grained (silt and clay-sized) material that was incorporated in turbidity currents. Because of its small grain size, it stayed in suspension far beyond the recognizeable edges of coarser-grained turbidite deposits. Other fine-grained clay minerals and clay-sized quartz is derived from continental sources, carried by wind until it drops into the ocean and slowly settles to the bottom. If the clay minerals from either of the above sources contain iron, they turn into what are known as "red clays" by the oxidation of iron in the clay as the clay settles slowly through the ocean water.

Another major constituent of abyssal plain sediment is calcium carbonate and silica that make up the skeletons of microscopic plants and animals that live in the ocean. When these micro-organisms die, their skeletal material sinks to the bottom, creating fine-grained material that is interlayered with the inorganic constituents. If the water depth is greater than about 5 km,

the calcium carbonate material dissolves, leaving silica as the only organic skeletal fraction. This depth is known as the **carbonate compensation depth.**

Still another major contributor to abyssal plain sediment is the volcanic ash that is produced by either submarine or subaerial volcanic eruptions. This material, as you might guess, is an important contribution in the vicinity of volcanic eruptions (mid-ocean ridges, island-arcs, etc.) and less important where these features are not present.

One final constituent is mineral material that actually forms on the ocean floor, at or just below the sediment-water interface. These minerals, usually clay minerals, are known as **authigenic** (formed in place). One particularly important authigenic deposit is known as a **manganese nodule.** These globular features are composed primarily of manganese and iron oxides and may be important ores of some concentrated minor metallic elements, like nickel and cobalt and copper.

Finally, occasionally a deep-ocean basin becomes cut off from normal circulation. If this happens, surface water of normal salinity can continue to flow into the partially closed basin, but deep water circulation is cut off (see Figure 13.13). Evaporation proceeds, the water

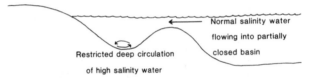

Figure 13.13. Partially closed basin in which normal salinity water can flow in, but higher salinity water has a restricted circulation and is unable to mix with normal salinity water.

becomes progressively saltier, denser and sinks to the bottom of the basin. Because deep water circulation is restricted, this deep water cannot mix with normal ocean water. If evaporation of the water in this (partially) closed basin exceeds the input of fresh (river) water, the water in the basin can become saturated in sea salts and begin precipitating evaporites. Apparently, this happened in the Mediterranean Ocean during the Miocene Epoch, leading to the formation of deep-sea evaporites.

Non-Marine Depositional Environments

By our definition, non-marine depositional environments include all depositional environments on the continental surfaces. These include: (1) **alluvial** (river) **environments,** (2) **lacustrine** (lake) **environments,** (3) **aeolian** (wind dominated) **environments** and (4) **glacial environments.**

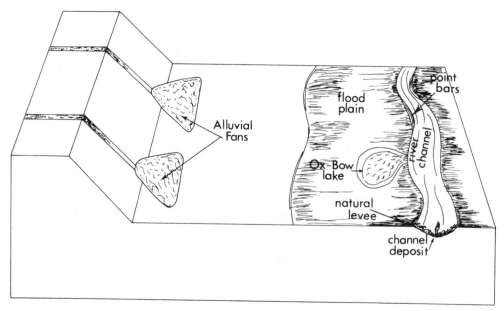

Figure 13.14. Block diagram showing possible relationships between alluvial fan, fluvial and flood-plain environments.

Alluvial Environments

Included in this section are several subenvironments: (1) alluvial fan subenvironments, (2) fluvial (river channel and meander loop) subenvironments and (3) flood-plain subenvironments. Figure 13.14 shows diagrammatically how these can be related to one another.

Alluvial fans are fan-shaped deposits often found at the base of mountains and often associated with arid climates. Alluvial fans have many of the same characteristics of turbidites discussed earlier. Both are the result of **sporadic deposition.** Both can contain sediment with a wide range of grain sizes. Both can exhibit graded bedding. Alluvial fans are deposited during periods of heavy rain in shallow lakes at the base of high-relief areas or subaerially. The coarsest material is deposited nearest the high-relief area; progressively finer material is deposited vertically on the coarse material and progressively farther from the source area.

Fluvial subenvironments include **river-channel deposits** and **point-bar deposits.** River-channel deposits are a function of stream velocity and turbulence, as discussed in the previous chapter. The maximum size particle a river can carry in its bed load is a function of river velocity. The amount of material and maximum-size particle that a river can carry in its suspended load is a function of river velocity and turbulence. As either factor decreases, some of the bed load can no longer be transported, and some of the suspended load is deposited. In general, channel deposits in straight stretches of a stream are coarsest near the headwaters of the stream and are progressively finer-grained downstream, reflecting a general decrease in velocity and turbulence from headwaters to the mouth.

Rarely, however, does a stream follow a straight path for a very long time. Most of you are probably aware that most rivers **meander** (excluding those in mountainous areas). The cause of river meandering is not known for certain. What *is* known, however, is that when rivers begin to meander, they continue to meander, with the meander curves becoming more and more pronounced. What else is known is that when rivers meander, some parts of the stream bank are eroded, while deposition takes place in other areas. The velocity of stream increases somewhat on the outside of the curve and decreases somewhat on the inside of the curve (analogous to two runners on a curved track—the runner on the outside has farther to travel and must run faster to keep up with the runner on the inside of the curve). As you might have guessed, then, the outside part of a meander curve is where the stream bank is eroded. Because the inside of the curve is subjected to slower velocity, some of the suspended matter in the stream is deposited, forming **point bars.** Point bars are usually characterized by finer-grained material than the nearby channel deposits. Point bars tend to build up toward the center of the stream as meandering progresses. Neither type of deposit is laterally very extensive. Furthermore, channel deposits tend not to be very thick. Channel deposits can be quite long, following the channel, but generally are not wide or very thick. Point bars are not very long but can be quite wide if they reflect deposition over a lengthy period of time and can be relatively thick.

Finally, let's consider flood-plain deposits. By definition, the floodplain is that part of the fluvial system that is covered only when the river overflows its banks. When the river, at flood stage, overflows its bank, the velocity of flood water decreases sharply (because it is no longer confined to the channel). As with other river deposits, this decrease in velocity results in the rapid deposition of the coarsest part of the suspended load, resulting in the formation of **natural levees,** those long, but not laterally extensive bars of sand that line many rivers. Finer material in the flood water is carried farther from the channel and is deposited as thin layers of silt and clay. As with some of the other depositional environments already discussed, flood-plain sedimentation is sporadic rather than continuous. It results in long bar-shaped deposits of coarser-grained material (levee deposits) that grade laterally into thin-layered, laterally extensive silt and clay deposits.

Lacustrine Depositional Environments

Lacustrine (lake) **depositional environments** make up only a small fraction of the total sediment deposited worldwide. This holds for the present, and probably this generalization holds for the geologic past. In spite of their limited occurrence, these deposits are important because many lake deposits are associated with a high degree of organic matter, which, under the appropriate conditions, can be transformed into oil or natural gas. Lacustrine environments can also be very complex. The principal criteria determining the type of sediment deposited in a lake apparently include: (1) local tectonic environment and (2) salinity of the water. Let's look at each of these separately.

The local tectonic setting governs how much sediment will be transported to the lake and what type of sediment. If the lake in question is adjacent to a mountain range or some other high-relief feature, a great deal of poorly sorted, angular immature sediment will accumulate at the lake shore and in the lake. If the adjacent area is not of high relief, the amount of sediment delivered will be less; there will be less very coarse material, and the material may be more mature (more time exposed to chemical weathering in the source area).

The salinity of the lake will govern what type of nonclastic material will be deposited. In saline lakes, we might expect evaporite minerals to be precipitated; in fresh water lakes, evaporites will not be precipitated, but carbonates may be precipitated. The salinity problem is more complex still. The salinity of some lakes changes with time—many lakes that begin as fresh water lakes evolve into saline lakes with time, and *vice versa*. Let's examine lake sediments in terms of water salinity.

Fresh Water Lake Deposits

Fresh-water lakes are most commonly found in temperate to tropical climates where fresh-water input (rainfall and runoff) exceed evaporation. In many of these lakes, calcium carbonate is present in solution near its saturation limit. Changes in temperature can lead to the inorganic precipitation of calcium carbonate and the deposition of fresh water limestone. Limestone can also be produced by biological activity. Many carbonate deposits will form near the shoreline where water is shallow and water temperatures can fluctuate with air temperature.

Some clastic deposition also takes place at or near the lakeshore. Large sediment-laden streams can deposit large amounts of clastic material, forming lake deltas. Much of this material will be reworked into typical shoreline deposits (beaches, etc.). Finer-grained clastic material will be formed in the deeper parts of the lakes.

Thus, fresh-water lake deposits can be a complex interfingering and interlayering of coarse clastics and carbonates at the lakeshore, progressively grading into finer-grained clastic material in the deeper parts of the lake.

Saline Lakes

Saline lakes, in general, are found in arid and semi-arid climates where evaporation exceeds fresh water input (rainfall and runoff). Water input comes from **perennial (permanent) streams** and **ephemeral streams** (those that only flow during high rainfall seasons) and by unchanneled sheet flow of water during high rainfall periods. In saline lakes, lakeshore deposits are principally clastic in nature. Depending on the relief of the adjacent areas, clastic sediment may range in grain size from coarse to fine, or from medium-grained to fine. Generally speaking, much clastic materal is deposited at the lakeshore in deltas or alluvial fans. Much of this clastic material gets reworked at the lakeshore, leaving only the coarsest material at the lakeshore. Away from the lakeshore, progressively finer-grained material will be deposited.

Saline lakes are also commonly the site of deposition of evaporite minerals, most of which are confined to the deeper parts of the saline lake. These deposits are interlayered with the finest-grained clastic material.

Saline lakes then commonly show clastic sedimentation at the lakeshore that grades into progressively fine-grained clastic material and interlayered evaporite deposits toward the center of the lake.

Glacial Deposits

Glacial deposits do not represent sedimentation in any particular depositional environment. Rather, they occur wherever continental or valley glaciers have dumped their sedimentary load. These deposits are typically very poorly sorted, and they are not laterally extensive. Glacial deposits take many forms, depending on how they were deposited. Most deposits of poorly suited glacial sediment are known as **moraines;** most of which occur at the terminus of the glacier (its farthest extent) when it starts to melt, forming **terminal moraines.** Some material is deposited as **lateral moraines** at the sides of glaciers.

Aeolian Deposits

Aeolian, or wind-derived, deposits are quantitatively important in arid desert regions and along shorelines. Dunes are the characteristic form of aeolian deposits. The sediment comprising the dune is generally fine-grained sand and is deposited in nonhorizontal layers known as **cross-bedding.** As you might guess, from previous knowledge of the Great Sahara Desert of Northern Africa, dune sands are an important feature on continental land masses. They are now, and have been in the past, quite extensive, and they represent an important contribution to the sedimentary record.

CHAPTER 14
Sedimentation in a Plate Tectonics Framework

In the previous chapter, we described the physical features of sediment typically deposited in major depositional environments, with no regard to the tectonic setting. To reiterate, the reason for this approach is that a given depositional environment is not unique to a given tectonic environment: It is possible to find, for example, turbidite deposits on both active and passive continental margins and in island-arc tectonic environments. There *will* be differences in the deposits, but the differences will be principally mineralogical or chemical, rather than physical.

In this chapter, we will switch gears and look at the type of sediment deposited in particular tectonic environments. In most cases, sedimentation in a tectonic environment is affected by the type of depositional environment. We will examine these tectonic environments in terms of the relative importance of the various depositional environments and look at how the tectonic environment produces certain unique characteristics in each depositional environment—characteristics which can distinguish similar depositional environments in different tectonic environments.

The tectonic environments to be discussed in this chapter include:

1. Island-arc/trench environments
2. Active continental margin environments
   Continental arc/trench
   Continental arc/interior
   Folded mountain belt
3. Continental interior
4. Passive continental margin
   Shoreline, continental shelf
   Slope
   Rise
5. Passive Ocean Basin Environment
   Rise/plains
   Plain
   Ridges
   Plain/arc

Figures 14.1 and 14.2 show diagrammatically the relationships between each of these tectonic environments.

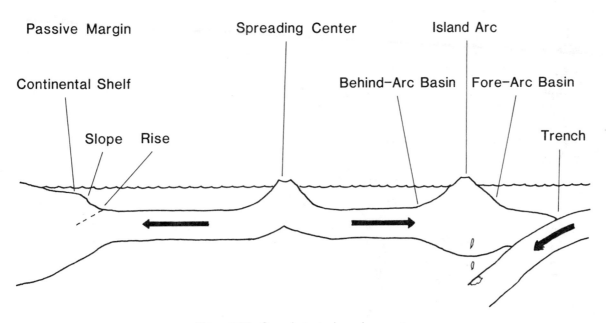

Figure 14.1. Oceanic tectonic environments.

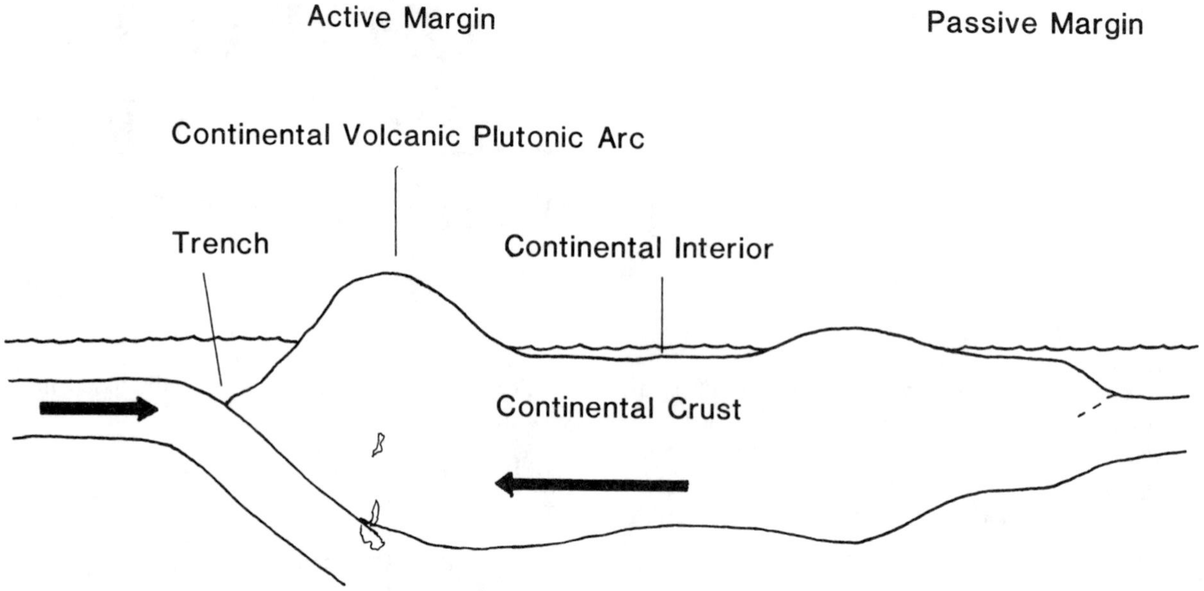

Active Margin

Passive Margin

Continental Volcanic Plutonic Arc

Trench

Continental Interior

Continental Crust

Figure 14.2. Continental tectonic environments.

Behind-Arc Basin

Island Arc

Fore-Arc Basin

Graywacke

Graywacke

Trench

Abyssal Plain
Sediments

Interlayered
Sediments

Graywacke

Figure 14.3. Sedimentation in the fore-arc and behind-arc basins associated with an island arc.

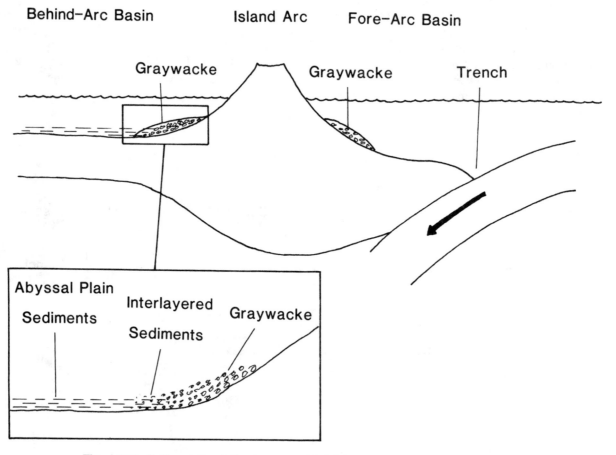

Island-Arc/Trench Tectonic Environment

In this section, we will examine sedimentation on the trench side of an island-arc tectonic environment. Figure 14.3 shows the various features of this environment that will be discussed. Most deposition in this tectonic environment takes place adjacent to the island-arc, which is an area of high relief relative to the ocean floor. The area is characterized by fairly dramatic change in slope at the base of the island-arc, where it changes from the moderately steep slope of the island-arc to the quite flat slope of the ocean floor. Because of these two factors, the most important depositional environment is the turbidite environment. On both the fore-arc side (nearest the trench) and behind-arc side of the volcanic-plutonic arc, immature, angular, rock-fragment dominated sediment accumulates on the steep flanks of the volcanoes under water, especially below the level at which waves can disturb this material. Periodically, these depositions slump, sometimes due to their deposition on steep slopes and resulting slope failure. At other times, the frequent earthquakes associated with this tectonic environment cause the material to slump. For whatever reason(s), this material begins to flow down the slope as a turbidity current. At the base of the slope, turbidites are deposited. This particular type of turbidite, containing much angular volcanic rock fragment is known as **graywacke.**

Because the arcs are almost always at least 100 km from the trench, most of the arc-derived sediment is deposited in the fore-arc basin, with very little finding its way to the trench proper.

On the other side of the arc, the behind-arc basin can grade into typical abyssal plain depositional environments or, if near a continental mass, into a marginal sea. In either case, deposition near the base of the arc is turbidite. Consider first the case of gradation into an abyssal plain environment. Away from the arc, we can find an area where the finer-grained material of the turbidite is *sporadically* deposited and abyssal plain deposition is essentially *continuous*. This results in an interfingering of deposits from these two environments, and the resulting sequence will grade from one to the other, both vertically and laterally. Figure 14.3 shows this relationship diagrammatically.

In the case of a behind-arc environment grading into a marginal sea, the same type of interfingering will occur, but the material interfingering with the fine-grained turbiditic material will be clastic material derived from continental sources and may be more like the sediment deposited in a continental-shelf depositional environment.

Although most sediment from the arcs is deposited in the fore-arc basin, some material does make its way to the trenches. The amount of sediment, however, varies from trench to trench. Some are almost filled; others are virtually empty. Due to the great depths of the trenches, geologists have little first-hand knowledge of the nature of the sediment. Undoubtedly, most of the sediment is fine-grained. Seismic studies indicate that what sediment is there is deposited in thin, flat-lying layers.

We have not discussed the nature of the sediment deposited in the near-shore environments—material deposited at the shorelines of emerged volcanoes. Because of the high relief and rapid erosion in these areas, coarse-grained conglomerates composed of angular material will be deposited at the shorelines. Compositionally, these conglomerates will be composed in large part of volcanic rock fragments and will be quite immature. Wave action will winnow the finer-grained material, the sands, silts and clays, which will, in turn, be deposited progressively farther from the shoreline. Because this tectonic environment is characterized by fairly steep slopes, much of the shoreline material will be subsequently eroded, redeposited in deeper water and ultimately end up as part of a turbidite. Thus, shoreline deposits in an island-arc tectonic environment must, to a large extent, be considered temporary features.

Finally, because subaerial erosion is generally rapid in an island-arc environment (high relief, much rain), it is possible to erode completely through the volcanic, or upper, part of the volcanic-plutonic arc, exposing the lower more coarse-grained plutonic igneous rocks. When this occurs, the nature of the sediment being delivered to the shoreline and ultimately to the fore-arc and behind-arc basins, changes. Rather than volcanic rock fragments, the sediment will be composed primarily of the large minerals of the plutonic rocks, quartz, feldspar and some type of iron/magnesium silicate. The sediment will still be immature and angular, but compositionally will be different from the material supplied by the volcanic igneous rocks.

Active Continental Margins

Active continental margins are those at which the edge of the continent is part of a collisional process—either a continent-ocean collision or a continent-continent collision. We will examine each of these separately, beginning with the ocean-continent collision.

Continental Arc/Trench Tectonic Environment

This is the tectonic environment at a continental-crust–oceanic-crust convergent boundary and is shown diagrammatically in Figure 14.2. Many of the environments will be quite similar to those previously discussed (island-arc/trench environments).

The continental volcanic-plutonic arc, in general, stands even higher than the island arc. This leads to the potential for even more rapid erosion of this boundary than at the oceanic-crust–oceanic-crust convergent boundary and the deposition of vastly larger quantities of sediment. Rapid erosion of the arc will yield angular, immature, volcanic fragments that will be deposited temporarily just below sea level, only to be retransported as turbidity currents and deposited as turbidite sediments in the fore-arc basin.

One difference between this and the previously discussed environment is that in this tectonic environment elevation of the volcanic-plutonic arc and continued compressional deformation, due to continued convergence, can cause parts of the fore-arc basin to be elevated above sea level. Exposure of the turbidites deposited early in the history of convergence can make them source rocks. Erosion of the first-cycle turbidites will yield sediment redeposited in second-cycle turbidites. Second-cycle turbidites will have many of the same characteristics of first-cycle turbidites (immature, angular, poorly sorted, graded beds), but some of the material shed by the exposed first-cycle turbidites could be more mature and better rounded due to more subaerial erosion and a second transportational history.

With more sediment being shed, more makes it to the trench in this tectonic environment, leading to thicker trench deposits. Most of the characteristics discussed previously will apply to trench deposits in this environment as well.

Continental Arc/Continental Interior Tectonic Environment

On the continental side of a continental volcanic-plutonic arc, immature, poorly sorted sediment being shed by the arc will be deposited in either a subaerial environment or in a shallow-water epeiric sea. In the case of subaerial deposition, the sediment will take the form of alluvial fans, thick deposits near the feet of mountains, not laterally very extensive. If deposition takes place in a shallow epeiric sea, the sediment will be more like that of the typical shallow marine depositional environment discussed in the previous chapter. In this case, we might expect to see fairly coarse-grained deposits (conglomerates) at the shoreline. Finer-grained material (sands, silts and clays), if deposited at the shoreline, will be winnowed by wave action. We might then expect to see progressively finer-grained material away from the shoreline.

The composition of the deposited sediment will depend upon the composition of the source rock (arc). In the early stages of erosion (and deposition), we might expect volcanic fragments to be an important fraction of the sediment delivered to the shoreline. Continued erosion will unroof the plutonic rocks of the arc, and the composition of the sediment will reflect that change. We might expect to see large grains of quartz, feldspar and an iron/magnesium mineral—still immature, still angular, but compositionally the sediment will be different. The sedimentary term applied to feldspar-rich sandstones is **arkose,** and the entire suite of sedimentary deposits on this side of the arc complex is known as **molasse.**

Folded Mountain Belt Tectonic Environment

This tectonic environment is created by the collision of two continental masses, yielding strongly deformed sedimentary and metamorphic rocks on both sides of a granitic core. Because folded mountain belts result from the collision of continental masses, the land surrounding the deformed area will be continental in nature. Many of the physical characteristics of the sediment deposited adjacent to this tectonic environment will be similar to those described just prior to this section, the sediment of the "continental-arc/continental-interior environment." Thick molasse deposits will also be common in this environment. Deposition can be either subaerial or in a shallow marine environment.

The composition of the sediment will reflect the composition of the source area, which will be different from the composition of the continental-arc/continental-interior molasse deposits. In this environment, the early sediment will probably reflect both the deformed sedimentary and metamorphic rocks which will serve as a "roof" over the granitic cores. The molasse will be composed of coarse-grained, poorly sorted, immature and angular fragments. The fragments will be composed of deformed sedimentary rocks and progressively higher grade regional metamorphic rocks. When this complex is completely eroded, unroofing the granitic core, then fragments of this granitic core will also be incorporated into the sediment being deposited. Because this tectonic environment generally produces little, if any, volcanic activity, volcanic rock fragments will be a very minor constituent.

Continental Interior Tectonic Environments

Strictly speaking, this is not an *active* tectonic environment. By definition, continental interiors are generally stable rather than tectonically active. We include them here simply to make our trek from active continental margins (first section) to passive continental margins (next section) complete.

Sedimentation on continental interiors includes alluvial fan deposits, all alluvial environment deposits (channel deposits, point bars, flood plains), all shoreline deposits and all shallow marine environment deposits, including transgressive, laterally extensive sandstones, siltstones and shales.

Shorelines adjacent to the high-relief volcanic-plutonic arc will tend to be characterized by coarse deposits (conglomerates) with progressively finer-grained clastic material away from shore. Other shorelines, not adjacent to high-relief areas, may not have conglomerates at the shorelines, but rather sandstones, with progressively finer-grained sediments away from the shoreline. Low-relief areas adjacent to shorelines will also yield *less* sediment and *more mature sediment* to the shoreline, because the source area will be subjected to a longer period of chemical weathering.

Some shorelines may not receive any (or very much) sediment, depending on the elevation and character of the land surface adjacent to it. In these cases, we might expect to find carbonate deposits near shore.

Finally, the sea level of epeiric seas rises and falls; when sea level falls, some continental interior basins may be cut off from free circulation of normal-salinity salt water. In such a situation, evaporation will lead to progressively more saline water in the closed basin and the eventual deposition of evaporites.

Passive Continental Margin Tectonic Environments

Again, these environments are not tectonically active. A passive continental margin is classically seen on the east coast of North America, where an extensive continental shelf is the site of clastic sediment deposition. The characteristics of the sediment deposited here will be identical to those described in the previous chapter under "Shallow Marine Environments— Shoreline and Offshore Clastics."

Shoreline deposits will be laterally variable, depending on the physiography of the shoreline. Some areas will be characterized by deltaic deposits where major rivers enter the ocean; other areas will be characterized by barrier islands, lagoons and marshes; still others will be characterized by estuaries—drowned river valleys at the shoreline. Finally, where clastic sediment is a minor component of the shoreline environment, carbonates will be deposited. The principal characteristic of shoreline deposits is that they will be laterally variable, grading into one another, not laterally continuous. Composition of the sediment will be a function of the composition and relief of the source area. Typically, any high-relief source area will be several hundred miles inland from the shoreline. In that case, the sediment being delivered to the shoreline will be fairly mature, well-rounded and medium to fine-grained. The same would hold true if no high-relief source area were providing sediment. Only if a high-relief source area were near the shoreline would we see an appreciable difference in the physical or chemical characteristics of the sediment delivered to the shoreline; in this case, we might expect to see more coarse-grained, immature and angular sediment deposited at or near the shoreline.

Offshore clastics will tend to be more laterally continuous and thin-bedded. The coarsest sediment, generally sandstones, would be deposited nearer the shoreline. Progressively further from the shoreline, the sediment being deposited would be finer-grained— sandstones, grading into siltstones, in turn, grading into shales.

Slope Deposits

Slope deposits on a passive margin will be quite similar to the slope deposits described in the previous chapter. On passive margins with extensively developed continental shelves, slope deposits will be fine-grained—silts and clays and will be quite mature. In general, the principal mineral constituents will be silt- and clay-sized quartz and clay minerals. Layers will be thin and will show evidence of deposition on a slope, including slumping and sliding.

Continental Rise

The continental rise on a passive margin is the boundary between the continental slope and the abyssal plain. This environment will be characterized by *turbidite* deposits—fan-shaped deposits from turbidity currents flowing down the slope and depositing their sediment at the base of the slope. The physical characteristics of turbidite deposits have been described in the previous chapter. Mineralogically, the sediment will be fairly mature, especially that portion that is derived from the continental shelf. Because turbidites are agents of erosion, they can pick up material eroded from the rock outcrops on the slope. This material will not undergo any appreciable chemical weathering and, therefore, the sediment derived from slope erosion could be mature or immature, depending on the mineralogical makeup of the slope rocks being eroded.

Finally, some reworking of turbidite deposits has recently been verified. Relatively swift cold water currents flow deep within the ocean. One particular type of current that flows roughly parallel to the shelf edge but deeper (near the junction of the rise and slope) is known as a **contour current.** Recent detailed studies of quartz grains in and adjacent to turbidite deposits suggest that these contour currents erode unconsolidated sediment in turbidite deposits, transport it laterally along the continental rise (see Figure 14.4) and deposit it whenever the contour current loses velocity. Much more work needs to be done to understand completely the various sedimentary processes that take place in this environment. Unfortunately, the depth of the water prohibits direct examination, except for an occasional (expensive) trip in one of the small submersible vehicles that are built to withstand the very high pressures.

Passive Deep Ocean Basin Environments

Passive deep ocean basins are inactive tectonic environments. Sediment deposited here is characteristically like that described in the Abyssal Plain depositional environment of the previous chapter. By and large, this sediment will be composed of thin-layered, laterally extensive beds of fine-grained clastics and some silica and/or carbonate material of biogenic origin.

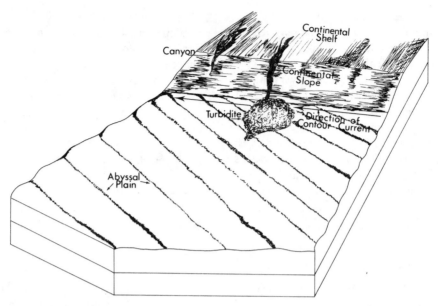

Figure 14.4. Block diagram showing deep-ocean contour current.

Near the continental rise, the abyssal plain sediments grade laterally into the farthest reaches of the turbidite deposits. If a spreading center is driving this passive deep ocean basin, sediment adjacent to the spreading center may contain appreciable volcanic ash from the volcanic activity related to the spreading center. Finally, if the passive deep ocean basin is adjacent to a volcanic-plutonic island arc, abyssal plain sediment on the deep ocean floor will grade laterally into the sediment of the behind-arc, described earlier in this chapter. Figure 14.1 shows diagrammatically the relationship between these various deep ocean tectonic environments.

Finally, where the volcanic ridge of a spreading center is below sea level, as it is along much of the Mid-Atlantic Ridge, for example, some sediment gets trapped in the medium rift valley of the volcanic ridge. This sediment will be composed primarily of fine-grained clastic material (silt and clay), some volcanic ash and some silica and/or carbonate of biogenic origin. We might expect a higher proportion of volcanic ash in this restricted environment relative to the open deep-ocean basins.

Summary

This chapter has *briefly* described the typical sedimentary deposits to be expected in various tectonic environments. The story appears very simple in that maturity, grain size, sorting, bedding and all sedimentary parameters are governed by physical and chemical processes that are active in the source area, during transportation and in the depositional environment. Complications arise because there is no simple "typical" sedimentary environment, and there is no "typical" sedimentary package for a given tectonic environment. Local differences in source rocks, length and direction of weathering, and transportation and depositional environments conspire to complicate the story. What we have attempted to do is to give you some idea of the processes that are active and to present an *ideal* view of sedimentation—the real view is much more complicated.

CHAPTER 15
Evolution of the Appalachians: A Case History

In the previous chapters, we have examined plate motions as if each one existed independently, with no connections between them. Lest anyone believe this, we offer this chapter to demonstrate that different plate motions are in fact, related to one another. The eminent geologist, J. Tuzo Wilson, a pioneer in the theory of plate tectonics has suggested a plate motion cycle that includes the following sequences of events:

1. Continental rifting due to appearance of a new spreading center;
2. Establishment of a narrow oceanic gulf, continued rifting;
3. Growth of rift to a wide ocean basin, continued rifting;
4. Ocean closing due to reversal of spreading and beginning of convergence;
5. Continent-continent collision, cessation of convergence.

Appropriately, this is known as the Wilson Cycle. Diagrammatically, a Wilson Cycle would look like Figure 15.1a through 15.1e. Now, it may very well be that the Wilson Cycle is simply a theoretical construct, and we may not be able to find anywhere a geological record of a complete Wilson Cycle. We can find, however, several examples of partial cycles that overlap, confirming the possibility of a complete cycle.

The Appalachian Mountain system of eastern North America, one of the most thoroughly studied folded mountain systems in the world, shows evidence of an almost complete Wilson Cycle. In this chapter, we will examine the evolution of the Appalachians through time. Much of this chapter is based on the work of Robert D. Hatcher, Jr., of the University of South Carolina, and we very much appreciate permission to use his research findings.

Before tracing the geological evolution of the Appalachians, let us first look at the surface features of the whole system. Figure 15.2 is an exaggerated topographic profile and simplified geologic cross-section across the Appalachians, showing the various physiographic provinces and simplified geologic descriptions of each. Rather than trying to describe the evolution of each physiographic province, we will look stepwise

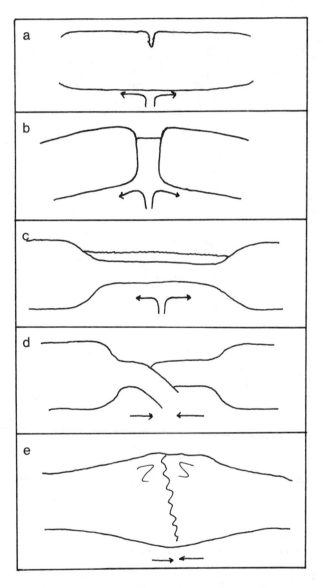

Figure 15.1. Wilson cycle. In "a" continental rifting is initiated; in "b" a narrow oceanic gulf is established; in "c" a wide ocean basin is established; in "d" ocean begins to close as divergence changes to convergence; in "e" there is a continent-continent collision.

93

Figure 15.2. Diagrammatic profile across the Appalachian Mountains. Rocks on the Plateau are Paleozoic sedimentary rocks, not folded but faulted along weak layers. Rocks in the Valley and Ridge are folded and faulted Paleozoic sedimentary rocks. Rocks in the Blue Ridge/Smokies are folded, faulted and metamorphosed. In general, rocks of the Smokies are younger Pre-Cambrian and not highly metamorphosed, while rocks of the Blue Ridge (Hayesville Thrust Sheet) are older Pre-Cambrian, folded, faulted and more highly metamorphosed. Rocks of the Piedmont are mostly metamorphic and igneous Paleozoic rocks. Rocks of the Coastal Plain are essentially flat-lying sedimentary rocks of Mesozoic and Cenozoic age, representing the eroded material from the folded Appalachians.

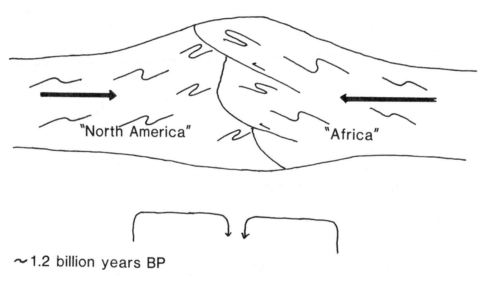

~1.2 billion years BP

Convergence – Continent–continent collision

Figure 15.3. Earlier convergence of "North American" and "African" continental masses, approximately 1.2 billion years BP.

at the evolution of the entire Appalachian system, describing how each event affected any or all the provinces.

Pre-Cambrian Rifting

Prior to about 820 million years ago, the continental masses that we now know of as North American, Europe and Africa were apparently one continental mass, perhaps the result of an even earlier continent-continent collision and folded mountain building event (see Figure 15.3.) Evidence for this very old convergence comes from the presence in the Blue Ridge of granitic rocks, now metamorphosed, that have been dated at more than 1 billion years. Approximately 820 million years ago, a spreading center beneath this continental mass began to split the mass apart (Event 1 of the Wilson Cycle). Continued rifting and spreading led to the development of an ancient ocean basin separating these continental masses (see Figure 15.4). This ocean has been named the **Iapetus Ocean.** Evidence of the ocean basin comes from the presence of Pre-Cambrian age basaltic rocks, now metamorphosed to amphibolites in the Blue Ridge province. Rifting continued until about 560 million years before present (BP), corresponding to Events 2 and 3 of the Wilson Cycle.

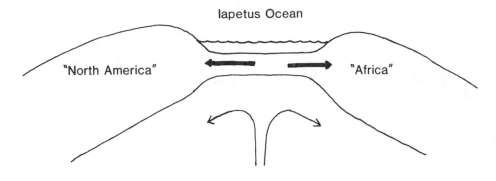

Iapetus Ocean

"North America"   "Africa"

820 – 560 million years BP

Divergence – Continent rifting and opening of Iapetus Ocean

Figure 15.4. Rifting of welded "North American"– "African" continental masses during the period 820–560 million years BP.

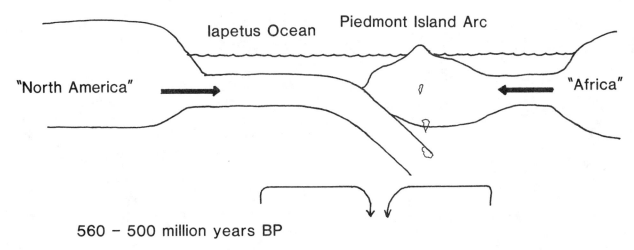

Iapetus Ocean   Piedmont Island Arc

"North America"   "Africa"

560 – 500 million years BP

Convergence – Closing of Iapetus Ocean, subduction

and formation of Piedmont Island Arc

Figure 15.5. Closing of Iapetus Ocean, start of convergence and establishment of Piedmont Island Arc, during the period 560–500 million years BP.

Convergence

Approximately 560 million years BP, whatever was causing the rifting and establishment of the Iapetus Ocean, essentially reversed itself, not only ceasing the rifting, but causing the continental masses to begin moving once again toward one another. This was probably initiated by the establishment of new spreading centers that started the continental masses moving toward one another.

Apparently, the new convergent plate boundary was located *within* the Iapetus Ocean. Although we are still unsure of some of the details, we now believe that the oceanic lithosphere of the North American plate was subducted beneath the oceanic lithosphere of the Af-

rican plate. The result, as we might predict, was the establishment of a volcanic island arc (within the Iapetus Ocean) which generated andesitic lava and immature graywacke-type sediment that began to accumulate on both the fore-arc and behind-arc areas (see Figure 15.5). For reasons that will become obvious later, this island arc is known to Appalachian geologists as the **Piedmont Island Arc.** An important point to keep in mind is that during this period, the Iapetus Ocean was closing, and apparently most of the closing was at the expense of the oceanic lithosphere of the North American plate, the one that was being subducted.

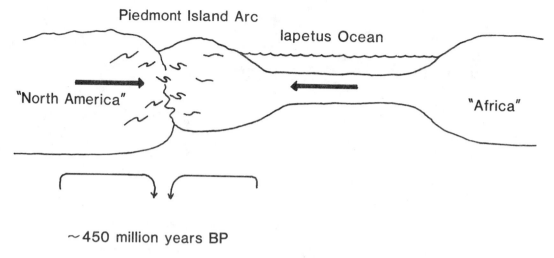

Piedmont Island Arc

Iapetus Ocean

"North America"

"Africa"

~450 million years BP

Convergence – Welding of Piedmont Island Arc

to "North America"

Figure 15.6. Welding of Piedmont Island Arc to "North America," approximately 450 million years BP.

Collision and Welding of Piedmont Island Arc to North American Continent: The Taconic Orogeny

During the Ordovician Period, about 450 million years BP, all of the oceanic lithosphere of the North American plate was subducted under the Piedmont Island Arc, which put the Piedmont Island Arc (now a major volcanic-plutonic arc complete with all of the typical associated sedimentary rocks) directly adjacent to the continental edge of the North American plate. Continued convergence caused a major collision between the North American plate continental margin and the Piedmont Island Arc (see Figure 15.6). This collision was essentially a continent-continent collision because the Piedmont Island Arc, dominantly andesitic in composition, had become a relatively thick **micro-continent,** unable to be subducted beneath North America or *vice versa.* This collision resulted in the welding of the Piedmont Island Arc micro-continent onto North America and is seen today as the **Piedmont province,** so extensively developed in North and South Carolina. Georgia and Virginia. Moreover, the forces responsible for the collision continued, causing severe deformation in both the North American continental margin lithosphere but especially in the Piedmont. The volcanic-sedimentary sequence shows both plastic and brittle deformation. Plastic deformation is seen in the many folds within the province and is most striking in the overturned folds, or nappes, of the province. Deformation was accompanied by regional progressive metamorphism of the volcanic-sedimentary rock sequence. Evidence of the highest degree of metamorphism is found with the most severe deformation and lower

degrees of metamorphism are found with the less severely deformed strata. Brittle deformation can be seen in the many faults of the Piedmont, now inactive. Some of the faults followed an extended period of plastic deformation, resulting in sheared-off limbs of nappes. This faulting probably occurred deep in the crust. Other thrust faults can be seen in less severely folded regions and probably represent a more shallow crustal deformation that included limited folding followed more quickly in time by thrust faulting.

A slightly more complicated model for this period of time has been proposed by Hatcher and Odom (1980). In this model, the rifting of Event 1 pulled from North America a small micro-continent that is supposed to have existed in the Iapetus Ocean. In this model convergence led first to the development of the Piedmont Island Arc also in the Iapetus Ocean (Event 4a), collision and welding of the rifted micro-continent back onto North America (Event 4b), followed by the welding of the Piedmont Island Arc onto North America (Event 4c). Figure 15.7 shows the sequence of events in this model. What we now call the Piedmont was originally two separate land masses that were welded onto North America at two discrete times, rather than a simpler model of welding an island arc onto continental North America.

Whichever of the two models described above is more accurate, we can say that the deformation that occurred during these events correlates to the thermal and deformational maximum in the evolution of the Appalachians. It was this deformational episode that produced most of the major structural features in the Piedmont and Blue Ridge—nappes in the Piedmont, and folding and thrust faulting in the Blue Ridge. It

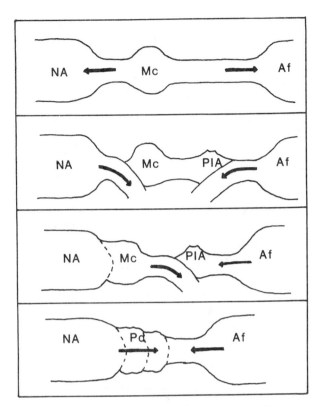

Figure 15.7. Alternative model for development of North American Piedmont. According to this model, rifting of "North America" and "Africa" (approximately 820 million years BP) caused small micro-continent (Mc) to be separated from "North America." Convergence led to welding of the micro-continent, followed by welding of Piedmont Island Arc, together forming the North American Piedmont. This was followed by collision of "North America" and "Africa."

was this episode that thrust the Hayesville Thrust Sheet containing metamorphosed Iapetus Ocean floor crust and sediment and even some underlying very old (greater than 1 billion years) gneisses that had developed in the earliest mountain building event, up and over young (late Pre-Cambrian) metasedimentary rocks (see Figure 15.2). It is generally believed that the Brevard Fault Zone of Georgia, South and North Carolina represents the zone along which the Hayesville Thrust Sheet moved up through the Inner Piedmont.

Deformation must have been extraordinarily intense, with both large-scale folding and large-scale faulting taking place. Probably both were happening simultaneously part of the time, faulting usually nearer the surface and folding at depth. Simultaneous faulting and folding may have also been taking place at the same depth, but at different geographic locations. Nearest the area where the collision was actually taking place and, therefore, where deformation was most intense, faulting was probably occurring; whereas farther away from the collision boundary, where deformation was less intense, the near-surface rocks were probably being folded.

Recognizing that both folding and faulting were taking place simultaneously, let us look at each separately, beginning with folding.

Specifically, let us examine folding in what is now the Piedmont of the Carolinas, Virginia and Georgia. Folding in the Piedmont ranges from relatively gentle, open folds in the eastern part of the Piedmont, to very tight, overturned folds in the western part of the Piedmont. The open, gentle folds of the eastern Piedmont are actually a series of variably intensely folded belts roughly parallel to the mountain belt.

One interpretation is that the open, gentle folds were formed nearest the surface and the tighter, more highly deformed folds were formed at great depths. Erosion of the gently folded strata in some places exposed the more highly deformed strata (see Figure 15.8).

Faulting in the Piedmont and Blue Ridge provinces is somewhat difficult to date accurately, perhaps partly because faults formed early (Event 4b) in this deformational period were reactivated several times during subsequent deformational episodes (Event 4c and Event 5.) What is reasonably certain, however, is that many major thrust faults originated in this deformation maximum.

The result of this episode, then, was a highly deformed folded mountain complex and an area of high relief. As noted earlier, when a land area stands higher than the surrounding area, the potential for shedding coarse-grained angular, immature sediment is also very high. This mountain complex began to shed such sediment into a shallow sea, generating sedimentary deposits, especially on the west or mainland side of the folded mountain complex. Clastic sedimentation continued throughout this Taconic deformation and the following periods of quiescence. The pattern of Paleozoic clastic sedimentation along the western margin of the folded mountain complex shows evidence of being generated predominantly by the continental area to the east. This sedimentary pattern is very similar to the ideal pattern of sediment deposition in a continental interior, in a shallow epicontinental sea with highlands on one side (the east) and low-lying land area on the other (the west).

Subduction Beneath North American Plate: The Acadian Orogeny

Approximately 370 million years ago, the Piedmont Island Arc was permanently welded onto the North American continent and severely deformed. Active volcanism had ceased about this time, the continued convergence of the North American plate and African plate took the form of an ocean-continent collision. Remember that the leading edge of the North American plate was the welded-on Piedmont volcanic-plutonic arc complex, with no leading-edge oceanic lithosphere. The

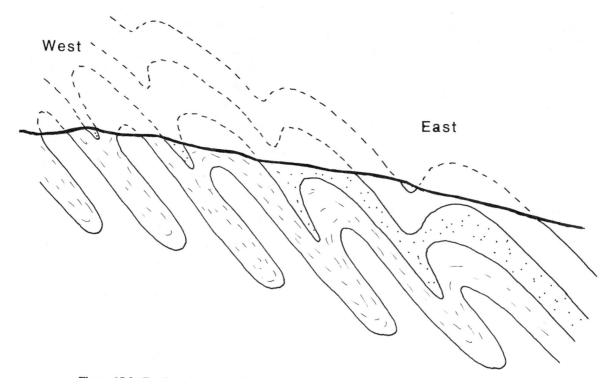

West

East

Figure 15.8. Erosion of gently folded Piedmont strata, exposing more highly deformed strata. Dashed lines represent material eroded away. Heavy solid line represents present ground surface.

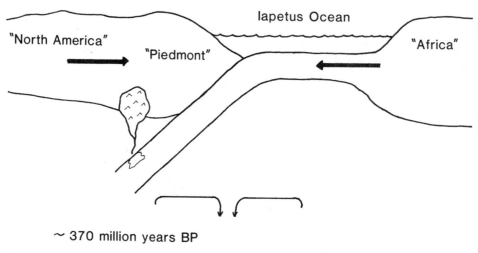

Iapetus Ocean

"North America"

"Piedmont"

"Africa"

~ 370 million years BP

Convergence – Subduction under "North America",

plutonic rocks emplaced

Figure 15.9. Subduction of "African" oceanic lithosphere under "North America," approximately 370 million years BP.

African plate on the other hand, had plenty of oceanic lithosphere on its leading edge. This type of collision, recall, leads to subduction of the oceanic lithosphere of one plate under the continental lithosphere of the other plate. In the evolution of the Appalachians, this is interpreted to mean a westward dipping subduction of African plate oceanic lithosphere under the North American plate (see Figure 15.9).

Recall from Chapter 10 that this type of convergence generally leads to the formation of nearshore trench, a continental volcanic-plutonic arc, and blueschist metamorphism. We may be in a little trouble here, in our interpretation of the evolution of the Appalachians. We seem to be missing two key ingredients in the rock record, andesitic volcanism and blueschist metamorphism of Devonian age. While we see no evidence of andesitic volcanic rocks of this age, we see *granodioritic* plutonism, in the form of the Spruce Pine (NC) granitic rocks which are dated at about 370 million years BP and which are intruded into the Blue Ridge metamorphosed basaltic and sedimentary rocks. These might represent the plutonic part of the volcanic-plutonic arc. Thrust faulting continued throughout this deformational episode. Most faulting was in the form of reactivation of previously formed thrust faults; however, apparently some new thrust faults appeared during this time.

Sediment would have continued to be deposited on the west side of the folded mountain complex in a shallow epicontinental sea, and we do find evidence of this. Of course, as the sea level continued to rise and fall, the shoreline migrated with time. Some areas were submerged most of the time and received an almost continuous influx of sediment. Other areas were submerged part of the time and above water part of the time. These areas show a sporadic pattern of sedimentation, when submerged, along with periods of no sediment accumulation, when above water. The type of sediment that accumulated at any place was determined to a large extent by how far that area was from the sediment source, and where the shoreline was at the time.

Collision of Continental North America And Continental Africa: The Hercynian Orogeny

The last stop in the collisional history of the Appalachians corresponds to Event 5 of the Wilson Cycle, the actual collision of continental North America (with the welded-on Piedmont Island Arc) with continental Africa (see Figure 15.10). This event began about 320 million years ago and continued for perhaps 80 million years. While the preceding plate interactions helped make the Blue Ridge and Piedmont provinces of the Appalachians into what they are, it was this event that shaped and developed the Valley and Ridge province of the Appalachian mountain belt, as well as affecting the previously deformed Piedmont and Blue River provinces. The location of the suture (where the continent-continent collision took place) is somewhat problematical, but is now generally believed to be somewhere in the eastern part of the Piedmont hidden under the cover of later Coastal Plain sedimentary rocks.

Deformation associated with this phase of the collision again must have been quite intense, as we see evidence of folds that formed during the Taconic deformational event, being *refolded* into very complex structures (refer back to Figure 8.9). We also see, but not as clearly, evidence that thrust faults that had formed during the Taconic and reactivated during the Acadian were reactivated once more. Also, several new thrust faults developed during this deformational phase.

In general, refolding and reactivation of thrust faults is recorded in the Blue Ridge and Piedmont. Some *new* structures developed in these areas. Farther west the sedimentary rocks that had been deposited in the epicontinental sea throughout the early and middle Paleozoic (Cambrian through Carboniferous) were deformed by being thrown into a series of overturned folds (refer back to Figure 8.8). These structures are among the newest in the entire deformational history. They affect Carboniferous rocks, meaning that this deformation must have taken place after deposition of these rocks. This allows us to determine roughly the age of this deformational event. Rocks as young as 350 million years old (Carboniferous) were folded; this indicates that the deformation took place afterward. We date the onset of this deformation at about 320 million years BP. The folded epicontinental sea Paleozoic sedimentary rocks became known structurally as the Valley and Ridge province.

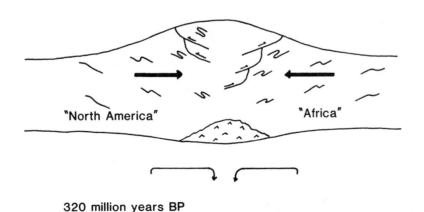

320 million years BP

Convergence – True continent–continent collision

Figure 15.10. Collision of "North America" and "Africa," approximately 320 million years BP.

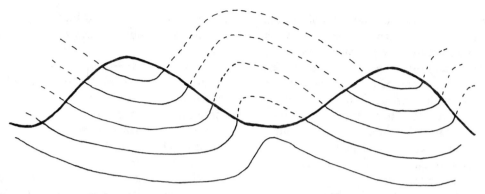

Figure 15.11. "Reverse topography" of the Valley and Ridge.

Interestingly enough, the present ridges are not generally where the anticlines were developed, as one might think. Rather, the ridges occur where the synclines were formed, resulting in what we term "reverse topography." The present day ridges and valleys are erosional remnants. Figure 15.11 shows diagrammatically how this topography developed.

The Valley and Ridge province also shows evidence of late stage thrust faulting as seen in Figure 15.2. Apparently, what happened was that because of the compressional deformation, the sedimentary rocks buckled into the Valley and Ridge folds. The compression was severe enough to cause some of the folded Valley and Ridge rocks and even some of the nonfolded rocks west of the Valley and Ridge to be thrust even farther west. Movement along these faults took place in the weakest rocks, generally shales. In general, stronger sediments such as sandstones and limestone were not fractured, but moved as a sequence along underlying shales.

Finally, recent deep seismic studies have suggested that the entire Piedmont may have been thrust west during this deformational episode. These studies suggest very shallow-dipping (almost horizontal) faults very deep under the Piedmont, in some places as deep as 15,000 feet below the present land surface. If these

faults are real, then the entire Piedmont as we know it today must have been thrust westward in much the same way as the more shallow thrust faults. Further studies have suggested that the rocks beneath this very deep thrust are still essentially horizontal, very different from the highly deformed, complexity folded Piedmont rocks. Much more work needs to be done to confirm this hypothesis.

Rifting of the Welded Continental Masses

The collision of continental North America and Africa during the Hercynian deformational stage resulted in a welded supercontinent containing what had previously been the continental masses of North America and Africa. The story does not end here, obviously, as today North America and Africa are separated by several thousand miles of Atlantic Ocean. Apparently, whatever caused the collision has reversed itself, causing the supercontinental mass to be split, resulting in today's continents of North America and Africa, separated by the Atlantic Ocean (see Figure 15.12).

Using several lines of evidence, we can date roughly the time when the plate tectonic forces reversed themselves. The first, and most direct line of evidence is the

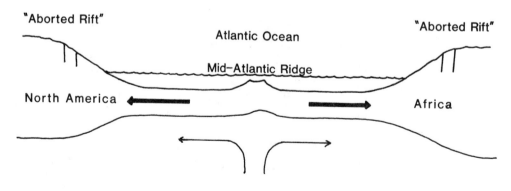

200 million years BP – Present

Divergence – Opening of Atlantic Ocean

Figure 15.12. Rifting of welded "North American"–"African" continental mass in two stages, initial rifting (approximately 170–190 million years BP) followed by spreading (begun approximately 200 million years BP.)

age of the present-day Atlantic Ocean. This ocean, remember, has formed by the rifting of the North America-Africa supercontinent. Therefore, the oldest part of this ocean floor, located immediately adjacent to the continental rises of both North America and Africa should reflect the age when the rifting began. The age of the oldest part of the Atlantic Ocean floor basalt, determined by radioactive age dating, is about 180 million years old. The Atlantic Ocean floor basalt has been found to be progressively younger toward the Mid-Atlantic Ridge, the present divergent plate boundary.

The second line of evidence comes from dating the sediment that has been deposited on top of the Atlantic Ocean floor. The thickness of sediment is greatest adjacent to either continental rise and is progressively less thick toward the ridge. Further, the age of the bottom sediment layer is oldest adjacent to the continental rises and is progressively younger toward the Ridge. The age of the oldest sediment is also about 180 million years.

The final line of evidence comes from the continents themselves. There exists all along the east coast of North America and the west coast of North Africa, a series of fractures in the crust which are filled with an igneous rock having the same composition (mineralogical and chemical) as basalt, but slightly more coarse-grained. These fractures filled with igneous rock material are known as **dikes,** and the basaltic rock filling the fracture is known as **diabase.** These **diabase dikes** have been dated radiometrically at about 170–190 million years old. These dikes apparently are an early "aborted" rift system along these coasts that began slightly before the "successful" rifting that resulted in the Mid-Atlantic Ridge. The fractures were formed by tensional forces. The fractures are much like those found associated with the Mid-Atlantic Ridge. Appar-

ently, the North America-Africa supercontinent began to split along what is now the east coast of North America and west coast of North Africa about 200 million years ago, but the rifting, for some reason, did not continue. Rather the ultimate rift zone (the Mid-Atlantic Ridge) formed several hundred miles away, at the edge of the present continental shelves, at a later time.

Finally, let's look at the erosion of the Appalachians, from the time of formation to the present. The Appalachian system, while still topographically considerably higher than the surrounding land area, is considerably lower at the present time than when formed. Constant erosion has worn the Appalachians down to the point where the Piedmont, perhaps once the highest part of the system, is now little more than rolling hills only several hundred feet above sea level, and the highest point in the Blue Ridge less than 7000 feet above sea level. For 200 million years, erosion has reduced the elevation of the mountain system, shedding sediment both to the west, most of which ultimately ends in the Mississippi River delta, and to the east where most of the sediment has been deposited on what is now a passive margin, trailing-edge sedimentary environment.

The sediment shed to the east has been mostly deposited at the coastline and offshore from the coastline. Looking at the east coast of North America now, especially along the Mid-Atlantic states down to Georgia, we can see a sequence of gently eastward-dipping sedimentary rocks, Cretaceous to Quaternary in age, known as the **Atlantic Coastal Plain** (see Figure 15.13). Most of these sediments were deposited at or near sea level, indicating that sea level has been several hundred feet higher in the past 8–100 million years than it is

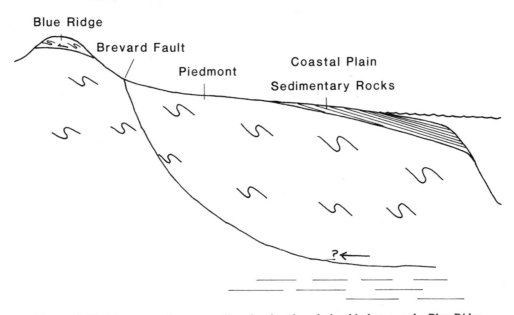

Figure 15.13. Diagrammatic cross-section showing the relationship between the Blue Ridge, Piedmont and Coastal Plain provinces.

now. The wedge of sedimentary rocks that form the Coastal Plain, continue onto the continental shelf, getting progressively thicker offshore. Some of the sediment drill cores taken from the continental shelf indicate that sea level has also been considerably lower in the past than it is now.

That takes us, in our story of the evolution of the Appalachians, up to the present. Can we predict what will happen in the future? Perhaps! We have no idea how long the North America plate will continue moving westward. As long as it does, the Appalachians will continue to be eroded lower and lower and the Coastal Plain sediment wedge will continue to grow thicker. We can predict, based on past history, that the North American plate will probably not continue indefinitely to move westward. History tells us that eventually the movement will reverse, the Atlantic Ocean will close, and a new mountain system will emerge. In that future time, we would probably see *geologic* evidence of this earlier mountain building event, but probably not *physiographic* evidence (the mountains would have eroded and become flat). The Coastal Plain sediments of the past 80 million years, now essentially flat-lying, will be caught up in the next collision and become deformed and metamorphosed much like the deformed rocks of the most recent collision.

Reference Cited

Hatcher, R. D. and A. L. Odom, 1980, Timing of thrusting in the Southern Appalachians, U.S.A.; Model for orogeny?: Journal of the Geological Society of London, vol. 137, p. 321–327.

CHAPTER 16
Oceans and Oceanic Processes

We have already seen that an understanding of the ocean's geological features is important to almost all explanations of major continental processes. However, we have not yet given much attention to the water which fills the ocean basins. In general, we know that the water is salty (about 3.5% "salt") and that it moves around, but not many of us are aware of the extent to which the water moves or just how much that motion can accomplish. As we look into this motion and its effects, we will start at the edge of the ocean, the coastline, and work our way out into deeper water.

Tides

Anyone who spends much time at the beach can testify to the fact that there is a lot going on along the coast. To most of us, the most obvious large-scale motion is the daily (or twice daily) rise and fall of the tide; likewise, the most effective agent of change appears to be the constant beating of the waves. Both of these processes seem rather simple on the surface, but become more complex when we investigate them in detail. Observations made thousands of years ago, for example, first indicated that there was a definite relationship between the position of the moon and the action of the tides. This permitted the early assumption that the moon is the primary producer of the tides. The reasoning went like this:

We can see that the moon appears to make a trip around the earth every 28 days, and we can understand how this works. The earth and the moon, each being a large mass of rock, have significant gravitational fields. Since the masses are only about 380,000 km (235,000 miles) apart, these fields pull the earth and the moon together rather strongly. They do not crash into each other, however, because the centrifugal force of the moon's orbit around the earth holds it away. Just as is the case throughout the universe, these two forces— gravity and centrifugal force—balance each other, and this balance has permitted the moon to maintain a reasonably stable orbit over millions of years. If this is the case, then the gravitational force which is holding these two bodies within a certain range of each other must be pulling rather strongly on their surfaces. We can begin to see a connection with the tides when we see that water, which is reasonably free to move around, will respond to this pull on the surface of the earth and

bulge out toward the moon, thus making a high tide on the side of the earth which is facing the moon. As the earth turns on its axis, each part of the earth's surface is carried into and out of that "bulge," making a high tide while in the pull, and a low tide while out of it.

So far, so good; but in fact most parts of the world have two high tides in each 24-hour period, not just one. We need to find some means to account for the second tide, and now the explanation gets a bit more complex. We have assumed that the moon is orbiting around the earth, and indeed, since the moon is rather small in relation to the earth, this seems quite plausible. If, on the other hand, we were to imagine that the earth and the moon were two planets of similar size that were close to each other, we would see that the only stable arrangement for them would be an orbit *around a point of balance between them.* In effect, they would have to orbit around each other much the way a baton tossed into the air must orbit around a point of balance between its weighted ends. There is no way that a freely moving baton could be made to spin such that one end remained in the center of a circle while the other end moved in a circular path around it.

Thus, in the case of the earth and the moon, it is necessary to see them as a "system" of two planets which rotate around their center of balance. Due to the great difference in size (the moon's diameter is about one quarter that of the earth), the center of balance between the two planets is located at a point within the earth, and thus the moon travels a great deal farther in its orbit than the earth does in its. Nevertheless, it is important to view the earth and the moon as a "system," since it is then easier for us to understand why each of them experiences the same forces. We have already mentioned the effect of gravitational attraction between the two planets. Now we can also examine the effect of complimentary centrifugal force.

As each of the bodies orbits around the center of balance, centrifugal force tends to pull it away from that center. We can see the effect which this force has on water if we remember, as children, taking a bucket of water and swinging it in a circle overhead. Centrifugal force pushed the water to the bottom of the bucket and held it there as long as the bucket continued to spin, even when it was completely upside down. In this way the mobile water on the earth's surface is "slung" to the outside of the circle as the earth and the moon

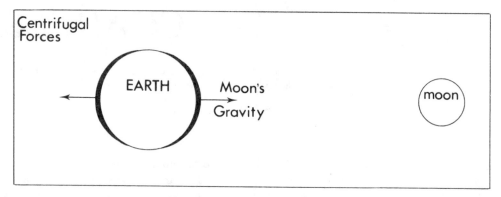

Figure 16.1. The two high tides produced by gravity and centrifugal force.

move around each other It is this phenomenon which produces a second high tide.

To summarize: We have a rotating earth-moon system which is held together by gravity and at the same time held apart by centrifugal force. The large bodies of water on the earth's surface respond to these two forces such that the water on the side of the earth facing the moon is pulled toward the moon by gravitation, and the water on the opposite side of the earth is pulled directly away by centrifugal force. This maintains two bulges in the earth's waters which we see as high tides. The areas in between, from which the water was pulled, we see as low tides. The rotation of the earth on its axis carries most parts of the earth's surface into and out of these bulges each day.

These are the basic means by which the tides are produced, but a careful observer of the tides would soon conclude that there must be more to it. An astute observer would see that the height of the high tide seems to vary each week, being higher than usual one week, lower than usual the next week, and so forth. Further, if this observer were to travel about, he/she would see that the daily tidal range (the difference between high and low tide) varies greatly from one place to the next. There are even some places which have only one high and one low tide each day. To account for these differences, we need to look at three additional factors which can affect the tides produced by the earth/moon system.

The first and most significant of these factors is the sun. It is possible to look at the earth and the sun in the same manner in which we have just considered the earth and the moon. The earth's orbit around the sun is controlled by the same gravitational and centrifugal forces which control the moon's orbit around the earth; in effect, the sun is as capable as the moon of producing tides on the earth, but because of its great distance from the earth the sun's tidal effect is only about 40% as strong as that of the moon. If for some reason the moon were to disappear completely, the earth would still have tides, but they would be smaller. So the tides that we observe are really a combination of two separate sets

of tides—the sun's and the moon's. When these two sets are aligned, during times when the sun, the earth and the moon form a line, the tides are higher than usual. (See Figure 16.2.) These are called "spring" tides, and occur twice each month. At other times the sun and the moon lie at an angle to each other and are trying to produce their tides in different places. These forces are counter-productive, and result in tides that are lower than usual, called "neap" tides. These spring and neap tides come in a predictable monthly cycle and together account for the variations in the height of the tides which are related to time.

Those variations in the tide which are related to location can be explained if we investigate two additional factors. The first of these relates to the moon's "ground position," or that spot on earth which lies directly under the moon. In this spot, the moon's gravitational pull is the greatest, and thus we can expect it to be the point where the tide is the highest. (In fact, several factors prevent this from being precisely true, but in general it is true that the moon's greatest effect is within the area which lies directly below it.) The moon's orbit varies, and the track of its ground position moves from 28.5° north of the equator to 28.5° south of it. Within this zone, the effect of the moon is going to be the greatest and the tides generally the largest. As we move north or south of the zone, the tides will become progressively less impressive until we reach the polar areas, which have practically no tides at all. Thus, if we are concerned with variations in tides in a specific location, we must first consider where that location is in relation to the ground position of the moon.

The second step as we investigate tidal variations in a specific location is to consider the physical characteristics of the location itself. There is a great deal to consider if we wish to be precise—the depth of the water, the contour of the bottom, the strength of the river currents, the wind, and a host of other factors may each have a small effect upon the tide. In many places, however, it is the shape of the coastline itself which has the greatest effect. Few coasts are absolutely smooth; most have a number of indentations which range in size

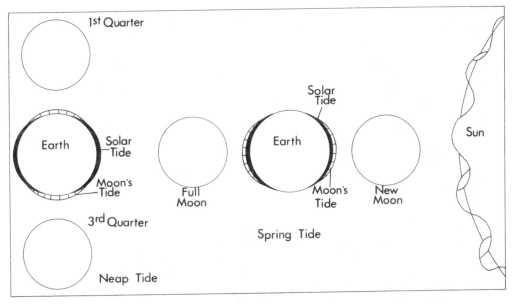

Figure 16.2. Interaction of solar and lunar tides.

from small bays to large seas. What is important is that each of these bodies of water responds differently to the rise and fall of the same tide. The reason for this is that each body of water has its own "resonant period," or rate at which the water freely "swings" back and forth.

In this respect, any body of water can be compared to a pendulum. A particular pendulum, once set in motion, will swing back and forth at a set rate, which is a function of the length of the pendulum. For example, if a pendulum has a shaft which is 1 meter long, it will swing with a period of just over 2 sec, no matter how hard it is pushed or how heavy it is. Similarly, a body of water will "slosh" back and forth in its container at a rate which is determined by the configuration of the container. If, in the case of the tides, a body of water has a natural period which is greatly different from the tide's twelve-hour period, they will work against each other and accomplish little, perhaps creating only one high tide per day. On the other hand, if the periods nearly match, they will work together like a parent pushing a child on a swing. Each "push" or "pull" of the tide will make the water swing higher in and out of the bay. The classic example of this situation is the Bay of Fundy along the eastern Canadian coast, where the bay's natural 12-hour period combines with its funnel-shape to amplify the ocean's natural 3-foot tides into great 40- to 50-foot tides at its head.

Thus we see that the combined effects of the earth, moon and sun can produce a significant rise and fall in the surface of the oceans each day. A great percentage of the oceans' water is involved in this tidal process, but the fact that the process occurs relatively slowly means that it is not capable of much geological work. In only a few places does the tide drive the water fast enough for it to pick up, carry and deposit any real load. The ocean's waves, however, are quite another matter.

Waves

In some places it is possible to see the waves accomplish a significant amount of geological work in only a few hours. Here again, the process seems quite simple at the start. Most waves are produced by wind. We have seen this happen many times. As soon as a small puff of wind starts across a pond, we can follow it by watching the ripples it is producing. We have all made our own ripples by blowing into our soup. In each case, the small waves are the result of the transfer of energy from the moving air to the surface layers of the water. Logically, the more energy that is transferred, the larger the resulting waves. Altogether there are three factors which control the amount of energy which can be passed on: The first is the **speed** of the wind. Obviously, high winds can cause larger waves than light winds; the second is the **length of time** the wind has been blowing—more time, more energy transfer; and finally, the third is the **area** over which the wind has been blowing. This is called the wind's "fetch" and refers to the fact that larger bodies of water can generate larger waves simply because they have more surface area to collect energy from the wind. This principle even accounts for the fact that the Pacific generates larger waves than the Atlantic because of its greater surface area.

Once produced, waves move away from their place of origin and can often travel for hundreds or even thousands of kilometers. This is possible because it is **energy** which is moving over that distance, not water. It is quite easy to demonstrate this process by floating a ping-pong ball on waves which are moving through an area free of wind. Following the motion of the ball, we find that it moves up and over each wave, then down the other side, returning nearly to its original location after the wave has passed. Each wave moves the ball

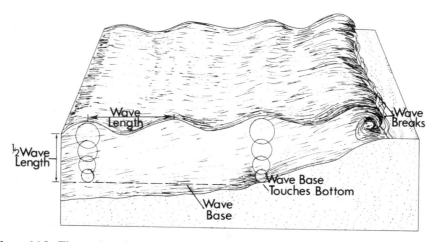

Figure 16.3. The motion of water within a wave that is moving from deep to shallow water.

around in a circle; since the ball is in contact with the water, it is evident that the water must be moving in the same way. (Waves represent energy passing along the surface of the water in much the same way that we can pass energy through a rug or a rope by shaking one end up and down. In each case, the energy passes through, but the transmitting material is not moved significantly from its original location.)

In the case of waves in deep water, this circular motion is confined largely to the upper layers. Observation indicates that the waves' motion decreases with depth and nearly dies out at a distance below the waves equal to one half of their length. Thus, if waves were found to be about 40 meters long (the distance from one crest to the next), we would know that they are disturbing the water to a depth of about 20 meters. This point, in effect the wave's bottom, is known as the **wave base.** Most waves in the ocean, even the very large ones, have a base no more than several tens of meters deep. Since most of the oceans are several thousands of meters deep, it may be seen that waves can pass fairly easily from one place to the next without a significant loss of energy. This means that the world's oceans become great collectors of energy which is steadily transported toward the coastlines. When the waves of energy reach the coast, they can accomplish a great deal of work by expending this energy over a small area in a short period of time.

As waves approach shallow water, they eventually reach a point at which the wave base is equal to, and then greater than, the water's depth. Under these conditions, the bottom begins to interfere with the circular motion of the waves. This uses up some of their energy and begins to slow them down. As the leading waves slow down, the following ones push in behind, and the result is that the waves are squeezed together. Pushed from behind, the waves can only react by growing larger and, in turn, becoming less stable. We can easily see this going on along any beach. As we watch the waves move in toward the shore, we see them start to rise

higher out of the water. At this time, they are being forced to slide along the bottom, thus slowing the lower portion of the waves more than the top. Obviously, this leads to a very unstable condition; the wave is getting higher and is being forced to tip forward as its foundation is pulled out from under it. In a short period of time, the top of the wave falls forward and it "breaks" onto the beach. It is at this point that the energy which the wave was carrying is transferred to the water and is expended as the water rushes up the beach.

Anyone who has stood in the breaking waves along a beach knows how much the waves can do. Each breaking wave can pick up a sizable load of sand, shells, or whatever is available and move it forcefully up the beach. If closely watched, this beach material will be seen to move in a number of directions:

1. First, and most obvious, is the motion up onto and down off the beach. It is clear that large waves can move more material than small ones, but it is also true that the large waves can carry the material for a longer period of time. This is important because it means that many large waves can hold their loads in suspension long enough to carry the sand not only onto the beach, but also back out again. The net result is that large waves tend to carry material off the beaches, while smaller waves with less transporting ability tend to build them up. This may be seen on many beaches which become smaller during the winter when more severe storms produce larger waves, then grow back when exposed to the smaller waves of summer.

2. The second motion is less obvious but equally important. The careful observer will note that most waves approach the shore at some angle, not straight-on. This gives the waves the ability to carry material along the beach as well as on and off of it. As a wave breaks on the beach at an angle, the water will wash up the beach at that same angle until its energy has been spent. Gravity will then take over, and the water will run straight down the slope of the beach back into the ocean. This means that the sand which that wave picked up

when it broke will eventually come to rest a certain distance down the beach. As each wave repeats this process, the face of the beach and the zone of surf just in front of it begin to act like parts of a stream in which a load is picked up, suspended for a time, and transported "downstream." This operation is known as **longshore transport,** and carries many millions of tons of sand and other beach material along coastlines every year. It is this process which is usually responsible for the cutting away of beaches in certain areas and the buildup of sand in others. It works just like a stream by eroding in areas where it has the energy to do so and depositing material in others where its energy is diminished.

Coasts

As one might expect, some coastlines are exposed to larger waves than others, and thus experience more erosion. The result is that coasts look very different from one place to the next. Compare the rocky coast of Maine or California with the sandy coasts of the Carolinas or the muddy coast of Louisiana, and one can see a complete spectrum from highest to lowest level of erosion by waves. At the lowest level, one finds areas in which the ocean's waves are so small and infrequent that they cannot bring about any significant erosion or transportation of coastal materials. In this very *low energy* environment, the beach is made up of whatever material is deposited there by processes on land. Generally, deltas and marshlands will be common here, since the ocean is incapable of moving any materials along the coast or out into deeper water. Such coastlines will usually be very irregular and muddy. Further along the spectrum of energy is the type of coastline which is exposed to waves of *moderate energy*. In this case, the processes of longshore transport work well most of the time. The very lightest materials (particles of silt and clay) are easily carried away, while the heavier sands and light gravels are spread out along the coast to form smooth, straight, sandy beaches. At the high energy end of the spectrum, the waves are large and able to cut away and transport almost any material. The result is usually a steep, rocky beach which lies at the foot of a cliff. This type of beach changes frequently as the waves cut away at the base of the cliff and advance the beach landward. Only the heaviest materials are left on the beach, while the rest are soon washed away.

The word "beach" was used frequently in the preceding discussion of wave action. The word is familiar to most people, but within geology it carries a specific definition. A **beach** is the area of a coastline which is affected by waves. This means that it extends from the highest point on land which is reached by storm waves out into the ocean to the point where the waves' bases first contact the bottom. (For the sake of definition, this point is said to be a depth of 10 meters.) Within this zone of abundant energy, some of the earth's most rapid geological changes take place.

Islands

Extending seaward from the beach is the continental shelf. This gently sloping extension of the continent varies in width from less than a kilometer along the continent's leading edge to several hundred kilometers along its trailing edge. At a water depth of approximately 200 meters, the slope increases slightly to become the continental slope. And this, in turn, extends to the continental rise and eventually the abyssal plain. The processes of sedimentation in these areas have already been discussed as have most of their major geological features, but a group of rather small features remains. These are islands. Geologically, islands have little significance, but people have always found them fascinating. Most islands are formed when a volcanic mountain on the sea floor builds itself up above sea level. As it does so, it moves from a calm environment into a relatively active one. Below sea level there are very few processes of weathering or erosion, but as we have just mentioned, the materials of an unprotected coastline can rapidly be broken up and carried away by waves. As a result, within geological periods of time, most islands are quite short-lived. While they exist, however, they may be seen to pass through two or perhaps, three stages.

1. Almost all islands begin as **high islands.** So called because they consist of one or more steep, active volcanoes, most islands in this group are growing, with each volcanic eruption, at a rate which exceeds the sea's ability to wear them away. Iceland and the big island of Hawaii are typical of this type. These, and others like them, will persist as long as their volcanoes remain active, although their time is limited. The plates on which they rest will eventually move them away from their sources of heat and magma, and the volcanoes will become inactive. Weathering and erosion will get the upper hand, and many of the islands will disappear.

2. In order to survive beyond the point when its volcano becomes inactive, an island must be located in a nearly tropical area, a place where corals can grow. Corals are small animals which build protective shells; some types form large colonies which can grow into reefs. For corals to succeed at reef-building, the water must be warm, shallow and clear: The coral animals need warmth, and the companion algae on which the corals depend for food and reef-building cement need sunlight, which can pass only through clear, shallow water. Where conditions are right, substantial coral reefs form around the slowly eroding island. The actions of wind, rain and flowing water will eventually level the central position of the island, but the reefs will

protect the lower portions from attack by waves. The waves do tear away parts of the reefs, but since they are made of living material, the damage can be repaired. It is interesting that even this destruction is beneficial to the island, since the debris from the reefs is usually washed in to form large beaches which significantly increase the island's land area. Islands such as these, surrounded and protected by coral, are variously called **low islands, coral island** or **atolls.** These islands will continue to exist as long as their protecting coral reefs can grow and repair themselves faster than the waves can tear them apart. To a certain extent, a coral reef can even adjust to new conditions, such as an increase in water depth, provided the changes are fairly slow. Rapid changes, however, can kill it. A recent example of this sort of change is the coming of the Pleistocene ice age. Rapidly falling sea levels killed off many reefs and exposed them to rapid erosion above sea level.

3. The few islands to survive this event fit into the third group, known as **aeolean islands.** These are named after Aeolus, the Greek god of wind, and they formed as the wind blew the coral debris from exposed reefs into great dunes which solidified to become hills. Some of the "hills" remained as the snow and ice began to melt and the seas began to rise. As the water rose once more, new coral reefs established themselves on the sides of the solidified dunes. Once again protected by coral reefs, some aeolean islands, such as Bermuda, are holding their own against the waves.

Currents

Most of us realize that there are many currents in the ocean which move great volumes of water from one place to the next. We may even be able to name a few of them such as the "Gulf stream" or the "California current," but most of us do not know why the currents move or what they accomplish.

It seems reasonable to divide the oceans' currents into two distinct sets for close examination. One set, that which includes the Gulf Stream, consists of discrete, fast-moving "streams" of warm or cold water which move across the oceans' surfaces. These "streams" are easy to find and to trace; for example, as early as 1786 Benjamin Franklin prepared a chart of the Gulf Stream as an aid to mail ships travelling from America to Europe. The other set of currents is not easy to see, but our knowledge of the ocean indicates that they must exist. These are large, vertical top-to-bottom currents that must exist to sustain the ocean's living things. We know, for example, that the ocean is filled with plants and animals whose needs must be met. The oceans' plants require a variety of nutrients along with abundant water and sun light. The animals, of course, must be supplied with oxygen. On land these substances are available in approximately the same place. The plants give oxygen off into the atmosphere where it is available to the animals, and both the plants and the animals deposit nutrient-rich wastes in the soil where they are available for re-use by plants. In the ocean, the situation is made more complex by the fact that water does not transmit light very well. (Only about 1% of the entering light makes it further than 150 meters down into the sea.) This means that most plants must live at, or near, the surface. They produce oxygen there, but most of their wastes, as well as all other material rich in nutrients, falls to the bottom. Animals can, and do, live throughout the ocean, but their only source of oxygen is the plant life at the surface. These two facts lead us to the conclusion that some large vertical, top-to-bottom mixing system must exist that carries oxygen down to the animals and brings nutrients back up to the plants. This must be the second set of currents.

Though harder to see directly, this second set is easier to understand, so we will look at it first. Large vertical motions in any fluid, water or air, are usually driven by variations in density. We can certainly see this principle operating in the air where cold and therefore more dense masses of air "sink" toward the ground, while less dense masses of hot air rise. The same process works in the ocean. Ocean water tends to be warmest, and thus the least dense, in equatorial areas where the earth receives the greatest amount of heat from the sun. The water is coldest, and the most dense, at the poles where the least heat is received. In the early 1800's, Alexander von Humboldt suggested that this situation could lead to the establishment of large convection currents which move water down toward the sea floor at the poles, along the bottom toward the equator, up at the equator, and finally back along the surface to the poles. Modern research shows the situation to be considerably more complex, but the net effect of this difference in temperature and density is to establish a circulation of this sort in the earth's major oceans. These convection currents move at a rate of approximately one revolution per 1000 years.

In a similar vein, it has been found that smaller bodies of water that are not subjected to great differences in temperature can still be stirred by currents based on differences in density. In these cases, the water can be made more dense by an increase in its content of "salt." The Mediterranean Sea is a prime example, for it lies in an arid area where evaporation exceeds precipitation. As the water in the Mediterranean gets saltier, it also gets heavier in relation to the water in the adjoining Atlantic. This results in the fact that the Mediterranean water tends to "sink" into the deeper parts of the Atlantic, and is replaced by water which is less salty. Within the Mediterranean, this leads to vertical mixing because the less salty water comes in on the surface, sinks as evaporation increases its salt content, and finally flows back out to the Atlantic along the bottom.

Currents of the type we have been describing are called **Thermohaline Currents,** indicating that they are driven by differences in temperature and salinity. Basically, they are density currents, and their prime effect is to bring about a thorough mixing of the oceans. This mixing brings oxygen down to the lower parts of the ocean, brings nutrients up to the surface, and serves to mix the oceans' salts so successfully that sea water contains exactly the same salts in exactly the same proportions everywhere on earth.

A consideration of the more obvious "stream" currents which flow along the surfaces of the oceans is more complex, because the motion of these currents is related to the motion of the wind. In order to understand their operation, we must first examine some of the actions of the atmosphere. We will begin by remembering that the atmosphere is subjected to the same differences in temperature as the ocean, and we may expect it to respond in the same way. That is to say, the very cold air at the poles should sink toward the earth's surface, while the much warmer air at the equator rises away from it. This in fact does occur, but the atmosphere is unable to establish the same worldwide circulation pattern which occurs in the ocean. The problem is that air is much lighter than water, and thus it can respond and move much more quickly. As it does so, its path is diverted by the rotation of the earth beneath it. This diversion is known as the **Coriolis effect,** and it operates on all objects which move over the earth's surface; the faster an object moves, the more obvious the effect. One way to explain this effect is to assume that we are watching the earth from a station on the moon. We pick two spots on the earth which we are going to track for a 24-hour period. One spot is at the north pole, and the other is on the equator. If we follow these two spots through one rotation of the earth, we find that the equatorial spot travels approximately 40,000 km (once around the earth). Since this is accomplished in 24 hours, it could be said that the spot is traveling at a speed of over 1,600 KPH (in relation to our observation post on the moon). In comparison, the spot at the pole merely turns once around and does not appear to move at all. We may generally assume that the air which is in contact with the earth's surface at each of these spots is moving in a similar manner. Thus, if air were to "try" to move from one of these spots to the other, it would be affected by the fact that its destination is moving at a different speed than its place of origin. We can compare this situation with that of the railroad passenger who tries to get off a train before it has come to a stop. There is little doubt that his course will be diverted by the fact that he is trying to move between two spots which are moving at different relative speeds.

We admit that this explanation is rather simplistic, but it should be an aid to understanding the factors which influence motion on a rotating earth. The net result of this influence is the Coriolis effect, which deflects all motions in the northern hemisphere to the right of their intended course, and all motions in the southern hemisphere to the left. Applying this effect to the circulation of the atmosphere, we see that the warm air that is rising at the equator and moving north will actually move to the right of its course and end up going to the east; similarly, the cold air which is moving south from the pole will have its course diverted so that it is eventually moving to the east. These diversions serve to break the atmosphere's circulation within each hemisphere into three sections. (We will use the northern hemisphere as an example. This same pattern would appear as a mirror image in the southern hemisphere.) The pattern consists of two strong circulation cells separated by a weaker zone. One of the strong cells is located in the polar area and is driven by the sinking of the very cold polar air. This cell covers the area from about 60° north up to the pole at 90° north. It is responsible for the winds which blow in that area, but it is not of great interest to us at the moment because there is not a great deal of open seawater at either of the poles. In this chapter, we will consider the other two cells, each of which has a significant effect on the ocean.

The other strong cell in the northern hemisphere originates at the equator and extends to about 30° north. Within this cell air rises off the equator and begins to move north. It gets to about 30° north before its course is diverted by the Coriolis effect, and it sinks back toward the surface. The majority of this air moves back along the surface toward the equator. This returning air is moving toward the south, but again it is diverted to the right, or to the west. (In usual terminology it becomes an easterly wind, because we describe a wind in relation to the direction *from* which it is coming.) These easterly and northeasterly winds become the strong and dependable "tradewinds," which blow almost constantly across the tropics and along the equator. For our purposes here, the prime effect of these winds is to eventually move the surface water very slowly in the same direction. This statement may appear to conflict with an earlier one which said that winds produce waves but do not actually move any water. The difference is a matter of time. The tradewinds, unlike most winds, blow in the same direction for years. As they do so, they produce waves, but eventually they also apply enough energy to the surface layers of the ocean to begin to move them along with the wind. This motion is slow in relation to the wind, but it produces a significant flow of surface water along the equator toward the west. This is known as the **Equatorial Current.**

Outside of this cell, in the area between 30° north and 60° north, lies the weak zone which is supplied with air which "leaks" out of the cell to its south. This air is heading north, but again the Coriolis effect diverts it to the right to become a west wind. It is not as strong or dependable as the tradewinds, but it is still able to

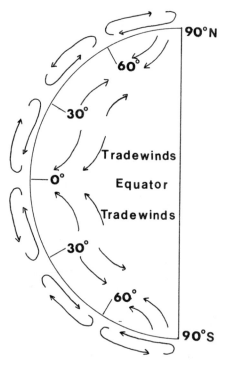

Figure 16.4. Major wind motions on a rotating earth.

move the surface water in the northern Atlantic and Pacific from the west to the east. When this flow is combined with the Equatorial Current which moves from the east to the west, we have the beginning of a clockwise circulation in the surface waters of these two oceans. The circulation cell is completed with the addition of north-south currents which connect to the ends of the two east-west currents. For example, the Gulf Stream in the North Atlantic carries warm water from the western end of the Equatorial Current up along the American coast to the northern part of the ocean. On the other side, the Canary Current carries cold water down along the European and African coasts to supply the Equatorial Current at its eastern end. A similar situation exists in the northern portion of the Pacific Ocean, and thus its surface water also moves around in a clockwise direction. In the oceans of the southern hemisphere, a similar arrangement exists, except that the circulation there moves counterclockwise.

The primary effect of this type of circulation is on the earth's climate. Water carries heat well, and thus these currents are able to improve the distribution of the earth's heat. The Gulf Stream, for example, serves to warm portions of northern Europe and produce milder climates there than would exist without it. Similarly, the cold California Current cools the San Francisco area and produces its famous fogs.

Summary

This chapter has concentrated on the oceans' water and looked at the forces which move it and the effects of that motion. Along the coastline, the primary effects are brought about by waves. We have seen that most coastlines are shaped by waves, and many are constantly being changed by them. In the open ocean, the water moves in a series of currents which serve to mix the water and distribute heat, oxygen and nutrients.

CHAPTER 17
Extraterrestrial Geology

For thousands of years, people have been fascinated by the many bodies that appear to move about in space above the Earth. We have named them, tried to map their courses, and speculated about their natures and environments. Until recently, however, we have had no way to test our ideas directly, and no way to collect "solid" evidence on which to build new theories. It may still be a long time before we get a close look at the stars, but the American and Soviet space programs have given us a good first look at several of our neighboring planets. For one thing, we now have better photographs of the planets than we have ever had before; in addition, we have the results of a variety of chemical and physical tests that were performed on-the-spot on Mars and Venus. Most important of all, of course, we have the results of actual human investigation on the Moon.

No doubt we will soon be able to send people to several other planets, and over the next decades we will be able to collect a great deal more information than we now have. Yet we will never know the other planets as well as we know the Earth; we will not be able to spend as much time examining them, nor will we be able to send as many investigators as have worked here. In the interests of efficiency we are going to have to take some shortcuts; we are going to have to use what we already know about the Earth to give us a head start elsewhere. To the extent that it is possible, we are going to have to use the Earth as an analog, and compare features we see on other planets to features and processes we already understand. Obviously, one must do this sort of thing carefully, for no other planet is going to be exactly like the Earth. At the same time, however, there are certain basic similarities on which we can depend. Our present evidence indicates that the elements that make up the solar system are similar throughout. Though the quantities and combinations may vary, the elements that make up the Earth's minerals are the same as those which form minerals on Mars, Venus or Pluto. Similarly, we can expect that the basic laws of chemistry and physics, which we understand, will apply everywhere. The strength of gravitational force may vary from one place to the next, but its effect on objects at a planet's surface can be expected to be the same.

In this chapter we will look at the Moon, Mars and Venus, those three planets which are closest to the Earth and probably the most similar to it. What we really want to know is just how similar they are.

The Moon

The Moon is the Earth's closest neighbor, and appears to be the largest object in the sky. Actually, the Moon is the smallest extraterrestrial body we can see with the unaided eye, but its proximity makes it appear large and allows us to see it and its features quite clearly.

We have long known that the Earth and the Moon are quite different, and it appears that the more we learn about the Moon, the more different it proves to be. Modern study of the Moon began with the work of Galileo in the early 1600's. At that time he noted that the lunar surface consists of large dark areas which he called "maria" (seas); these are surrounded by lighter-colored areas he referred to as "terrae" (lands). This early analogy was soon discarded when we learned that the Moon's mass is only about 1/80th that of the Earth, and thus its gravitational field is too weak to prevent gas and water from escaping out into space. If, however, the maria cannot be seas, what is it that they *can* be? Questions such as this became frequent as more and more sophisticated optical equipment allowed us to get better and better looks at the Moon. Eventually, telescopes became good enough to let us see any object on the Moon larger than a football stadium. Spacecraft let us look even closer, and finally the Apollo missions allowed a few of us to actually walk on the Moon's surface. All told, fifty spacecraft have come close to the Moon or landed on it, twelve men have explored its surface, and 380 kilograms of rock and soil have been returned to the Earth for study. As a result, we know a great deal more than we used to, but we still have a great many questions.

The Moon's Surface Features

In addition to the large and obvious maria and terrae, the predominent features on the lunar surface are **craters.** The closer one looks at the Moon, the more craters one can see, for they come in all possible sizes. The largest craters were easily seen by Galileo and measure over 200 kilometers in diameter, and the smallest craters can be seen only with a microscope; there are all sizes in between. It would appear that the formation of craters is one of the primary processes shaping the surface of the Moon. As we attempt to explain the craters' origin, we can look at the ways by which craters form on Earth. The most common cra-

Figure 17.1. Impact craters in a lunar mare (sea). Irregular, light-colored terrae (lands) at edge of picture. Light-colored rays surround major crater. (Photograph courtesy NASA.)

ters here are the result of volcanic activity. Some are formed by explosive eruptions; others are the result of the collapse of surface material into a void left by the extraction of magma. Less common are craters such as Meteor Crater in Arizona, which are the result of the impact of meteorites on the Earth's surface. Until we visited the Moon, it was difficult to determine which of these processes was responsible for the craters we saw. A good case could be made for either possibility.

Recent exploration of the lunar surface indicates that there are craters of both varieties, but the majority are the result of impact. This conclusion is based on the finding that the debris surrounding most craters is older than the material on which it is lying, and thus it is likely that it was placed there by the explosion which follows the impact of objects traveling at high speeds. This view is further supported by the existance of **rays,** which are light streaks arranged in a radiating pattern around many craters. The rays mark spots where debris from the crater disturbs the lunar soil.

Once the origin of the craters is understood, it is possible to return to the maria and the terrae. The round shape of the maria, and their great size, suggests that they might also be products of impact. Closer examination shows that, in fact, they are produced by both volcanic and impact processes. Apparently, the mare is

first excavated by the impact of a great meteorite. This event releases a great deal of heat and simultaneously releases the pressure on the rocks at the bottom of the crater. Increased temperature and decreased pressure lead to melting, which sometimes produces enough lava to fill the crater and occasionally enough to spill out over the surrounding terrain. These flows appear darker than the surrounding terrae because they are younger and have not yet been as extensively broken up by impact craters.

Running between and occasionally through the craters, maria and terrae are a great number of cracks and sinuous lines. These are called **rilles** and obviously fit into at least two categories. The first type are the straight lines, or **linear rilles,** which are best explained as faults. They appear to cut across most other features, and in some places it is clear that one side is higher or lower than the other. A few linear rilles appear in pairs to form **grabens,** and most interestingly, some serve to localize minor volcanic activity and thus create lines of small volcanic craters.

The second type of rilles, the **sinuous rilles,** were the subject of more controversy because their curving path slightly resembled the course of a stream. Had not this explanation been ruled out by the Moon's lack of water, it would still have seemed unlikely because the rilles

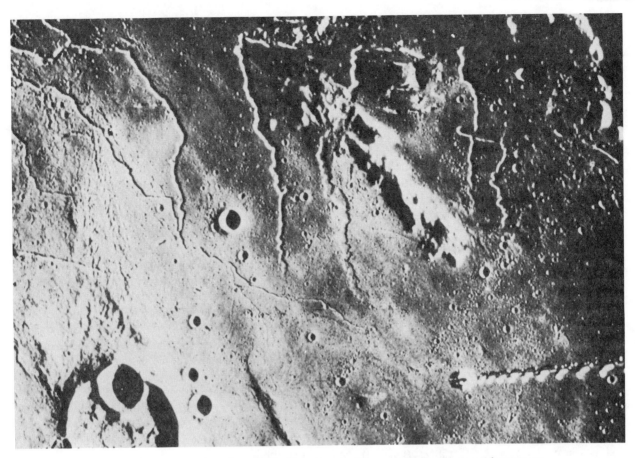

Figure 17.2. Lunar rilles. Screw-like object is part of the spacecraft. (Photograph courtesy NASA.)

all had blunt ends instead of the dendritic pattern which we associate with streams. In this case, the best analog was a feature on Earth called a lava tube, which forms as lava flows down a mountain. The lava on the outside and top of the flow cools in contact with the air and solidifies to form a tunnel or tube through which the rest of the lava flows. When the eruption ends, the liquid flows out of the tube and frequently parts of the structure collapse. Looked at from above, these partially collapsed tubes are long sinuous valleys with abrupt beginnings and ends. This theory is supported by samples of basalt collected by the crew of Apollo 15 which had landed beside Hadley Rille. This feature resembles its earthly counterparts in all ways except for size. Hadley Rille is almost a kilometer across, and it extends for over twenty kilometers across the lunar surface. Lava tubes on Earth tend to be considerably smaller.

The final major feature on the Moon's surface is the **lunar soil.** At the inception of the Apollo program, the Moon's soil became a source of concern, because it was necessary to know how deep it was and how well it could support the weight of a man and a spacecraft. To solve this problem, it was necessary to know of what the soil consisted. Obviously, it could not be the same as Earth's soil, for most of the weathering processes which pro-

duce our soil do not operate in the absence of water. There remained two possibilities: In one case, the soil could be made up of very fine-grained "dust" which the Moon had slowly collected from space. In the second case, the soil could be pulverized rock produced either by the impact of meteorites or the heat-induced breakage of surface rock. Once we had access to samples, we found that the soil actually consisted of a variety of materials. The primary constituents are fragments of rock, but these are mixed with small droplets of glass and a sampling of the dust which is constantly being blown out from the Sun. This sampling of soil, in fact, proved to be one of the most efficient methods by which to study the Moon, for each sample of soil contained fragments which had been transported from several areas of the surface.

The Moon's Interior

We have known for quite a while that the Moon's overall density (3.3 gm/cc) is about equal to the density of basalt (3.0 gm/cc). When we discovered that the majority of the lunar surface consisted of basalt in various forms, it became apparent that the planet was not as dramatically stratified as the Earth. In an attempt to understand what lies below the surface, the astronauts drilled several holes into the Moon's crust and deployed a network of seismographs. The holes were used to determine the rate at which heat is being conducted out from the Moon's center; the seismographs recorded any seismic activity which might occur. With these two kinds of information, it was possible for scientists to estimate the composition and physical state of the lunar interior. The results indicate that the Moon's interior is hot, but not nearly as hot as the inside of the Earth. It is doubtful that any of the Moon's interior is liquid. It might be most accurate to describe the Moon's interior structure as mostly lithosphere and asthenosphere. The lithosphere appears to be quite thick and rigid; it extends to a depth of about 1000 kilometers, or halfway to the center of the planet. The majority of the remainder is probably a "soft" material similar to the Earth's asthenosphere. There may be a small solid iron core at the center. This theoretical model is supported by the fact that most lunar quakes occur at a depth of about 1000 kilometers, marking the boundary between the lithosphere and the asthenosphere. In addition, data concerning heat flow indicate that the Moon's internal temperature is probably lower than 2000°C. This is considerably lower than the temperature in the Earth's mantle, and less than half the estimated temperature in the core of the Earth. Information such as this indicates that the Moon is nearly a "dead" planet. Its internal temperatures are too low and its lithosphere is too thick for any sort of plate motion to be possible. This, in turn, rules out most types of change which could be brought about within the Moon and on its exterior. This does not mean that the Moon is absolutely static, but it does indicate that most of what occurs there must be initiated by some outside force. For example, it was found that the seismic activity on the Moon is produced by the "tidal" forces exerted there by the Earth. What quakes do occur on the Moon all occur when the Moon and the Earth are closest together, the times when the Earth's gravitational pull on the Moon is the greatest.

Further information concerning the Moon's inactivity comes from the age of the rocks picked up on the lunar surface. Some of the first rocks recovered proved to be 3.6 billion years old. Later information has shown us that this figure is about the average for lunar rocks. The range in their ages extends from 3.1 billion to 4.6 billion years old. You should compare this with the rocks on the Earth's surface, which are rarely more than a few hundred million years old. On Earth where the surface is quite active, the continental areas are constantly recycled through a series of constructions (orogenies) and destructions (erosion); similarly, the rocks which form the sea floors are continually created at spreading zones and destroyed in trenches. In contrast, the ancient lunar rocks have obviously remained near their place of origin for the majority of the planet's lifetime. The rilles and the flood-basalts in the maria are

evidence of an active period in the Moon's past, but since no igneous rocks younger than 3.0 billion years have been found, we must assume that the Moon's active days are long past.

Mars

We have always hoped that Mars would turn out to be very similar to the Earth, and we have found or created evidence to support that hope. Measurements indicate that Mars is considerably smaller than the Earth. It has a mass that is about 1/10th of ours and a gravitational pull slightly over 1/3rd as great. On the positive side, Mars rotates on its axis at almost the same speed as Earth (its day contains 24 1/2 hours), it has noticeable polar ice caps, and its axis is inclined to the plane of its orbit, producing an Earth-like sequence of "seasons." It has even been noted that the planet's reddish color seems to change slightly from one season to the next. Could this indicate the presence of vegetation? Many of us have hoped so, and in the past there were some who even believed they could see the famous "canals," which they claimed to be evidence for the existence of intelligent life.

We began to get a more complete and realistic view of Mars in late 1971, when the Mariner 9 spacecraft entered an orbit around the planet and began to photograph its surface.

Surface Features of Mars

Early photographs, taken by spacecraft which preceded Mariner 9, had shown a large number of Moon-like craters. These pictures lead many to fear that Mars might turn out to be just as "dead" as the Moon. The pictures from Mariner 9 indeed contained a number of craters, but they also provided strong evidence for a varied and active planetary surface.

The exterior of Mars is divided into two distinct regions which correspond roughly to its northern and southern hemispheres. The southern hemisphere appears to be the older of the two, for it is much more extensively cratered. Its surface is very rough and littered by debris of all sizes ejected from the craters. Near the boundary with the northern hemisphere, the terrain becomes more chaotic, and is disturbed by a number of cracks, fractures, rifts and depressions. Among these features, whose origins are not yet well understood, is one of Mars' most impressive features. This is an enormous rift valley named Valles Marineris, which is over 3,000 kilometers long, several hundred kilometers wide, and up to 6 kilometers deep. Many times larger than the Grand Canyon, this valley could extend almost all the way across North America.

The northern hemisphere of Mars is the site of most of its volcanic activity. This area contains a number of shield volcanoes as well as several lava domes, cinder

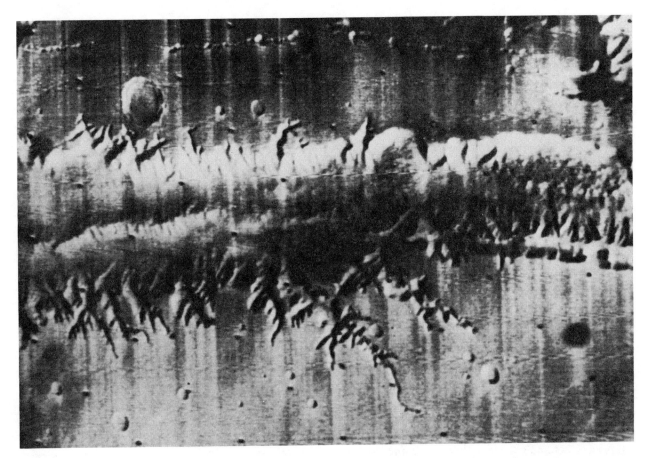

Figure 17.3. Valles Marineris. (Photograph courtesy NASA.)

cones and lava flows. Of the volcanoes, the most impressive is Olympus Mons, which may well be the largest shield volcano in the solar system. The mountain is 27 kilometers high (2.5 times as high as Mount Everest) and its base is 600 kilometers across. Its shape, and that of others like it, leaves little doubt as to its origin. The presence on its flanks of lava flows that appear to be recent suggests that it is still active.

Features in both hemispheres show the effects of erosion by wind and water. The effects of wind were immediately obvious when Mariner 9 approached, since it arrived during a planet-wide dust storm. The dust obscured the surface for several weeks. Researchers were surprised to learn that the atmosphere producing this storm was very thin. It proved to be less than 1% as thick as Earth's atmosphere, despite the fact that calculations had indicated that Mars had a sufficient gravitational pull to hold an atmosphere 10 times as great as that which exists. The atmosphere consists almost entirely of carbon dioxide (95%), with smaller amounts of nitrogen, argon and water. It has been determined that at speeds of approximately 200 kph, this small amount of gas can carry a significant load. In the southern hemisphere there are several large dune fields, which are evidence of the wind's transportational ability. As the wind's direction changes with the seasons, the orientation of these dunes changes; it is likely that

the orientation of the dunes would determine their ability to reflect the sun's light, and this might well account for the "seasonal changes" we thought we saw earlier.

The presence and actions of water on the Martian surface are more difficult to explain. Under present conditions of low atmospheric pressure and low temperature, any water on the surface would either freeze or evaporate. However, a large number of channels in the walls and on the floor of Valles Marineris appear to be identical to features on Earth which have been cut by flowing water. (See Figure 17.3.) Unlike the case on the Moon, there does not appear to be any other way to account for the formation of these "stream valleys." Similar features can be seen leading away from some of the depressions in the "chaotic" areas of the southern hemisphere. At present, the only way to explain this is to assume that the Martian atmosphere must have been thicker and warmer at some time in the past. This would have allowed liquid water to exist at the surface and cut the valleys. If, however, there was water available in the past, where is it now? And for that matter, where is the rest of the atmosphere?

Most researchers believe that both water and carbon dioxide are trapped within the present polar ice caps, where temperatures are quite low enough to form both water ice and "dry" ice. In fact, clouds have been noted at the polar areas, which may well mark the condensation of both water and carbon dioxide. Additional water is probably trapped below the surface in other regions, as a sort of "permafrost." The depressions in the chaotic regions may be seen as evidence of the existence of such permafrost. It is believed that the frost can melt and produce water which flows out through the "stream valleys"; such removal of water from the ground would allow the surface to collapse, forming a depression.

Structure and Composition of Mars

The overall density of Mars is about 4 gm/cc in contrast to about 5.5 gm/cc for the Earth. Mars is considerably smaller, however, and thus its interior is under less pressure. If this factor is taken into account, the "uncompressed" density of each planet is about the same, and it is therefore possible that their internal compositions are similar. Chemical analysis of the Martian surface materials supports this possibility in that the rocks there appear to be basalts. Certainly other similarities can be found in the existence of the shield volcanoes, the rift valleys, and the two distinctly different terrains. Unfortunately, we do not find any long trenches which might be zones of convergence, and we do not see any spreading zones or large transform faults. Olympus Mons provides good evidence for the fact that Mars has a hot and at least slightly active interior, but at the same time its size is strong evidence for the fact that Mars must have a very thick and immobile lithosphere; if the lithosphere were moving, it would probably have carried the mountain away from its source of magma before it could have grown so large. Similarly, if the outer covering of Mars were thin enough to break and move easily, it would not be strong enough to support such a great weight (even though the mountain weighs considerably less on Mars than it would here).

At present it appears that Mars is a much more "lively" planet than the Moon in that it is capable of bringing about significant internal and external changes under its own power. At the same time, it appears to be much more stable than the Earth. The majority of the Martian crust is probably just about as old as the Moon's surface. On Mars there are changes going on, but one change occurs on top of the previous one. If this is true, it will probably be quite easy to sort out the history of Mars some day, by simply digging down into the surface and reading off the events as they occurred.

Venus

With the exception of the Moon, Venus is the brightest object in the night sky. This is due to the fact that it is completely covered by a thick layer of white clouds which reflect the light from the Sun. Since the time of Galileo, these clouds have obstructed most attempts to collect more than the most basic information about the planet. For example, it was possible to determine its mass, size, density, and distance from the Sun. In relation to these data, Venus resembles the Earth more strongly than any of the other planets. More recent and precise information, however, points up some differences. The American Pioneer Venus orbiter compiled an extensive map of the planet's surface based on data from a radar altimeter. The map shows that the Venutian surface is quite smooth and uniform in comparison to the Earth. Venus has some large pleateau-like highlands (one of these is almost the size of North America), and there are also some volcanic mountains and some low areas; but the majority of the planet (70%) is covered by rolling plains. Unlike Mars and the Earth, the surface of Venus does not appear to divide itself into distinct regions. Most all of the surface appears to be one large "continent" which has been cracked, bent, and perhaps folded in several locations.

Data from the Soviet series of Venera spacecraft seem to support this view of the Venutian surface. These craft landed on the surface and sent back several pictures as well as some chemical and physical data. Venera 8 collected information which suggested that the crust in its area was composed of a material similar to granite. Venera 9 and 10 landed near a complex of shield-like volcanoes and indicated that they were probably sitting on basalt.

To most people, the most remarkable difference between Earth and Venus is found when comparing their atmospheres. On Venus the atmosphere is 95% carbon dioxide and about one hundred times as thick as Earth's. The "greenhouse" effect of the CO_2 holds the surface temperature at about 450°C (870°F). This unpleasant situation is made even worse by the fact that the only liquid present in these hot clouds is sulfuric acid.

A great deal more information will be needed before we can make any reasonable guesses about the interior of Venus, or try to map the history which must have led up to the present situation. Currently, the best estimates are that Venus began its lifetime very much like the Earth. The existence of the present atmosphere is evidence of a period of segregation in which its lighter materials were brought to the surface, and presumably heavier materials settled toward the core. The apparently extensive granitic crust might well have been generated at the same time, through a process similar to the Earth's current plate motion. Some investigators believe that this process was so successful that it eventually choked itself with an overabundance of light crust which could not be subducted. In any event, there is little evidence of any plate motion today. The area below the crust, however, appears to remain sufficiently active to produce some volcanism, and some sort of mantle "plume" (or hot spot) mechanism which supports the large highland plateaus.

116

The current atmosphere on Venus is thought to be the result of a lack of available water. The Earth has as much CO_2 at its surface as Venus has, but our atmosphere is very different because the majority of our CO_2 is bound up in limestones. Since water is an essential medium for the formation of carbonates, it is quite likely that Earth would resemble Venus more closely if our surface temperatures had been high enough to evaporate most of our water.

Summary

In this chapter we have attempted to compare the Earth to its nearest neighbors in the solar system. To the disappointment of many, we find that these neighbors probably resemble each other more strongly than they resemble Earth. In all cases, the Earth has been shown to be a more active planet. It is certainly the only one on which the process of plate tectonics is currently active, and with the possible exception of Venus, the only one on which the process ever had a significant effect. Since this process is so important to most geologic events, we are tempted to conclude that without plate motion the other planets must be much "simpler" and less exciting places. If we are to be realistic, however, we must merely say that they appear to be quite different, and we look forward to their further exploration, which will provide answers to many of our questions and give detail and color to the histories of our neighbors.

CHAPTER 18
Economic Geology

Economic geology is the science that concerns itself with the search for valuable materials within the earth's crust. Based on this very general definition, it could be considered the oldest of the geological sciences, since humans have been searching for a variety of valuable earth-materials for well over a million years. This search of ours probably began when hunters and foragers first looked for simple tools and weapons. It is likely that these early searchers discovered all too soon that the materials they wanted were hard to find. Though their tools and weapons were made of stone, only certain types of stone would suffice; they needed a substance that would break with a concoidal fracture and thus produce a sharp edge. No doubt they had to search through a great deal of unwanted rock in order to locate one piece they could use. At that time, the best advice to the searcher must have been, "Try to remember where you found it last time, and look for another place which appears to be similar."

With time, our needs have changed and our methods of search have become more sophisticated, but the basic problem and our approach to it have changed little. The growth of industry has brought demands for a greater number of raw materials such as manganese, sulpher, sodium and titanium. Our only source for such materials is the earth's crust, but all of these materials together represent only a very small percentage of the crust. As mentioned earlier, a chemical analysis of the crust reveals that it is composed predominantly of the elements oxygen (46%) and silicon (27%). If we add aluminum (8%), iron (5%) and calcium (4%), we have accounted for 90% of the crust's material with just five elements. All the other naturally-occurring elements, about 78 in all, make up the remaining 10%, and most of them account for far less than 1/10th of 1% each. This makes it quite clear that if we are going to locate these rare materals, and if we are going to be able to afford to remove them from the ground, we need to find places where natural processes have already done some of the collecting and concentrating for us. We are going to have to search for ores, naturally occurring concentrations of minerals which can be extracted at an affordable price.

With an increasing understanding of geology, we can guide our search for ores with the advice, "Find out which geologic process will concentrate the material you want, and look for a place where that process is (or has been) operating." In the broadest sense, the two geological processes which concentrate valuable materials are (1) igneous activities, and (2) the actions of weathering and erosion. We will look at these individually.

Igneous Processes

Our understanding of the origin of many metallic ore deposits took a great step forward in the latter part of 1978 and the early part of 1979. At that time, a group of geologists and oceanographers was investigating a segment of the Galapagos Rift, a spreading zone in the east Pacific. Their intention was to study the means by which heat is transferred from the hot, newly formed sea floor to the adjacent sea water. What they found was a far more effective transportation system than they had expected. They discovered several "plumes" of extremely hot water (several hundred degrees Celsius) rising directly out of the sea floor near the rift-zone. These features came to be called "chimneys" because the hot water's heavy load of suspended materials gave it the appearance of a dense cloud of smoke. The "chimney" effect was strengthened by the buildup of a small cone of precipitated material at the base of each "plume."

Later investigation showed that these chimneys originate when cracks form in the cooling flows of basalt at spreading zones. The cracks fill with water, and as they propagate, the water is able to find its way from one crack to the next. As the water moves, it is heated, and its eventual high temperature, combined with its original content of salt, allows it to become chemically active. A chemical exchange is promoted between the rock and the water, in which the rock gives up metals such as iron, copper, zinc and lead, and accepts in their place other metals such as magnesium, sodium and potassium. The result is that the water's "salt" mixture changes considerably, while the basalt is altered to a greenschist.

Eventually, the water's high temperature forces it to rise out of the rock, and it comes up as one of the plumes. As it does so, it comes into contact with the cold sea water and its physical and chemical conditions change rapidly. This change forces much of the hot water's dissolved load out of solution. Some of the load is precipitated immediately to form the solid base of the chimney, while the rest is temporarily held in suspension and forms the cloud of "smoke." In time, the fine particles in this cloud settle to form a reddish-brown mud which is rich in metals. Ocean drilling projects have encountered this mud in several areas of the Pacific and Atlantic, and many investigators believe that it forms a common bottom layer of sediment in many parts of the ocean. Measurements indicate that the mud varies in thickness from one meter to about ten. It consists predominantly of iron and manganese with smaller amounts of many other metals such as copper, tin and zinc.

Further, there is reason to believe that much of the material collected by these hot circulating waters is never delivered to the surface of the sea floor. We have already mentioned that some of the sea water's constituents are deposited in the basalt; others, such as sulfur, are believed to combine with the newly liberated metals and produce good-sized sulfide deposits below the sea floor. If they exist, deposits such as these would be too deep to be located by present drilling techniques. Our only evidence for their existence is found on the island of Cyprus in the Mediterranean. The name Cyprus comes from the Greek word for copper, and is altogether appropriate, since the island has been the site of copper mining for well over 4000 years. These mines provide valuable information because the island itself is a "wrinkle" of sea floor which has been exposed above sea level. The mines penetrate a layer representing the metal-rich muds, and extend downward into sulfide deposits which lie below.

Cyprus, of course, represents a unique situation. Most of the metallic deposits, in or on sea floor, remain buried under thousands of meters of conventional sediment and move with the plate toward a convergent boundary. At such a boundary, a portion of the sea floor's load of sediment is scraped off against the opposing plate, while the remainder descends with the plate into an area of increasing temperature and pressure. The sediment and the upper portions of the plate have adjusted considerably to surface conditions, and they each contain a great deal of water. This situation allows them to melt more easily than the material which surrounds them, producing magmas with abundant water and frequently heavy loads of metals. As these magmas work their way toward the surface, they can create one or more of the following types of deposits.

Disseminated Deposits

These deposits, often called porphyries, are formed as hot fluids, rich in metals, are boiled out of the sediments. These fluids move into and alter the country rocks which lie above the descending plate. The result is a large mass of quartz-rich rock containing metallic sulfides which are fairly evenly distributed. Deposits such as this form a band which extends from the southern part of the Andes Mountains, northward through Central America and up into the Rockies. In certain places, the sulfides are sufficiently concentrated to be considered ores, and may be the sites of large mining operations.

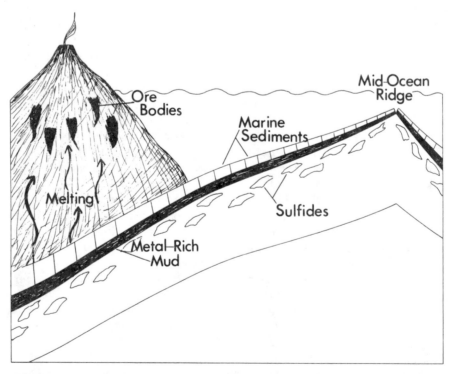

Figure 18.1. Metallic deposits placed in and on the plate are moved toward a subduction zone.

Gravity Segregations

These deposits originate in large magma chambers. In this case, the magma is trapped on its way to the surface. As it slowly cools, it passes through the crystallization temperatures of a wide variety of minerals. If all the necessary ingredients of any particular mineral are present in the melt when that mineral's crystallization point is reached, small crystals will begin to form. Since these crystals are denser than the melt, they will settle to the bottom of the magma chamber and collect there. In effect, the mineral in question will be collected from throughout the magma, and concentrated in a rather small area at its base. Such concentrations can be frequently rich enough to be mined.

Hydrothermal Deposits

These are by far the most common, and the most important, of the three types. As the name suggests, they are deposits from "hot water." They differ from disseminated deposits in that they are emplaced in cracks within the cool surrounding rock, and have little effect upon the rock itself. The fluid which forms these deposits can be produced by the heating of rock or the cooling of magma. As we know, when rock is heated to the melting point, the first liquid to be produced is often rich in water, quartz, and a variety of metals. This is particularly apt to occur at convergent boundaries where the original materials contain these components in abundance. Similarly, when a melt of this variety crystallizes, hot water with a heavy dissolved load of quartz and metals is often left over. As one plate moves below another, promoting a great deal of heating and cooling, a lot of "hot water" will be produced, and much of it will carry a rich dissolved load.

As this water works its way into fracture systems and finds its way to the surface, its temperature will go down, its pH will go up (it will become less acid), and it will be forced to deposit some of the minerals that it is carrying in solution. The predominant deposit will be quartz, which fills most of the cracks, but along with the quartz there will often be significant deposits of a great variety of metals. Silver, gold, lead zinc and many others are often found as deposits in these **veins** of quartz. Like the other deposits we have mentioned, hydrothermal deposits are commonly associated with mountain ranges and can be found throughout the major ranges of North and South America.

Weathering Processes

In many cases, deposits such as those above are quite rich enough to be successfully mined. In other cases, the metals are brought near the surface, but are not yet sufficiently concentrated to be considered ores. In the latter case, the many processes of weathering and erosion to which the deposits will be subjected may serve to concentrate certain valuable substances. There are three basic processes which can accomplish this: the first is weathering itself; the second is the transportation of the weathered products; and the third is the deposition of those products.

Weathering—"Residual Concentrations"

Much weathering occurs because the metallic components of minerals are more soluble in mildly acid surface waters than are nonmetals. As these metals are dissolved and carried away, the minerals are altered to sand and clay. Sand, which consists predominantly of silicon and oxygen, is not usually considered to be economically valuable, but clay can become important because it contains aluminum. Under conditions of extreme chemical weathering, such as in tropical areas, most of a rock's original material can be dissolved away, leaving the insoluble aluminum behind as a major component of the resulting soil. Under ideal conditions, this process can form **bauxite,** an ore of aluminum. Gold, which is even more resistant to chemical weathering, can form deposits in a similar way. Gold often originates as small flecks which are distributed throughout another substance. As the host material weathers away, the gold is left behind. Each of these cases is an example of **residual concentration,** a process which forms an ore by removing most of the unwanted material by weathering.

Transportation—"Mechanical Concentration"

We already know that flowing water tends to "sort out" materials as it transports them, heavy items being carried more slowly and for shorter distances than light ones. In the case of gold again, and several other materials, this process can produce ore deposits. In addition to being highly resistant to chemical weathering, gold is also very dense, and this makes it difficult for streams to transport. When gold particles are introduced into a stream, they will be carried only under conditions of the highest energy. If the stream's energy decreases, as it would with a change in gradient or at an obstruction, the gold particles will be among the first items to be deposited. If this deposition continues for a while in the same location, the gold can accumulate to form a **placer deposit,** a collection of dense material which has been concentrated by the action of flowing water.

Deposition—"Secondary Enrichment"

This process is the reverse of residual concentration in that the valuable material is the one which is moved. Waters flowing in and on the earth's surface usually collect a significant load of dissolved materal which they transport until changing conditions force the load out of solution. In the case of some low-grade metal deposits, the processes of solution and redeposition can

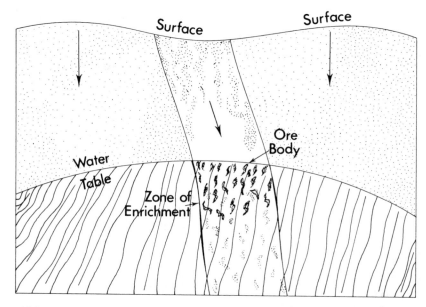

Figure 18.2. An ore accumulates through secondary enrichment at the top of the water table.

occur very close to each other. Metals picked up near the surface can be forced out of solution a few meters below the surface at the interface with the water table. If this process of collection and deposition occurs in a restricted area over a long period of time, it can produce a **secondary ore** which is considerably richer than the deposit from which the metal was collected.

Deposition—"Evaporites"

Some continental waters carry their dissolved loads until they empty into a lake or the ocean. This load accumulates for a while as the salts in the ocean or in certain salt lakes, but as we know from Chapter 12, much of this salt will eventually be deposited on the bottom of the body of water. One way by which this occurs is through evaporation. As the sea or lake water evaporates, its dissolved constituents become increasingly concentrated until some of them are forced out of solution as **evaporite deposits.** We already know that halite (NaCl) and limestone ($CaCO_3$) can be formed this way, but they are often accompanied by more valuable compounds from which we can extract potassium, phosphorus, and sulfur.

Energy

Up to this point, we have looked at processes which can form ores within the earth's crust, but economic geology must also consider other materials in and below the crust which are valuable because of the energy which can be extracted from them. Presently, the United States derives the majority of its energy from petroleum (oil and natural gas), a lesser amount from coal, and the smallest amount from nuclear fission and hydroelectric sources.

Petroleum

Chemical analysis of various oils, as well as geological examination of the rocks in which the oils originated, indicates that most petroleum forms from the debris of marine organisms. If this is correct, two conditions must be met in order to create a significant quantity of petroleum. First, there must be an abundance of organisms; second, their remains must be sealed off from the environment so that they do not become a source of food for other animals. The present delta of the Mississippi river provides one example of how this might occur. The river's water provides the necessary nutrients for the organisms in the Gulf of Mexico, and simultaneously it carries a great quantity of sediment which can bury and preserve the organisms' remains. In the past, similar conditions must have existed on a much larger scale along both convergent and divergent plate boundaries. The middle-eastern oil fields are an example of the former; they mark the ancient closing of the Tethyan Sea which used to separate Africa from Asia. The present and potential oil fields in the eastern and western Atlantic are examples of the latter use. Developers hope that these fields will draw on petroleum which accumulated when the Atlantic first began to open.

Once buried, the organic debris is subjected to increased heat and pressure. These conditions allow the organic material to break down and reform into a series of chainlike hydrocarbons. These are compounds, composed primarily of carbon and hydrogen, which range from methane (CH_4), which is "natural gas," to much longer and more complex chains. All of these are fluids, and as they are less dense than the water which fills the pore spaces in the surrounding rocks, they generally move out of their **source rock** and gradually rise toward the surface. It is believed that most of the earth's petroleum has escaped this way, and that only a fraction of one percent has been trapped in a **reservoir rock** on its way up to the surface. A reservoir, or an oil "pool," can be created when the petroleum moves through a permeable layer of rock and encounters an impermeable fault or a "dome-like" structure with an impermeable cap. Oil companies search for such "traps" and drill into them with the hope of extracting some of the petroleum.

Coal

The origin of coal is less mysterious. People have been finding and burning coal for over a thousand years, and in the process they have noticed that lumps of coal frequently contain the imprint of sticks, roots and leaves. This provides good evidence that coal forms from the remains of plants. Further investigation shows that the best place for these remains to accumulate is at the bottom of a swamp. Swamps are ideal because their abundant supply of water promotes the growth of plants, while the fact that the water is shallow and stagnant discourages the proliferation of the minute organisms that would consume the plants' remains. Shallow water cannot hold very much oxygen, and stagnant water can neither replenish its supply of oxygen nor carry away waste products. The result is that the plant material that falls into such areas is beyond the reach of most organisms, and is thus preserved.

Plant materials consist predominantly of carbon, hydrogen, oxygen, and a variety of other minor components. Once buried in the swamp, the oxygen is removed first and then other components are "squeezed" out as the pressure from overlying materials increases. The carbon is left behind, and thus its percentage of the total increases as the process continues. The first stage in this process is the formation of **peat,** which is a brown, almost pudding-like material which can be dried and used as a fuel. It is easy to light, but it burns with a smoky flame and does not produce a great deal of heat. Increased pressure leads to the formation of **lignite,** which is dark brown and solid. It is not considered a "true" coal, but it contains about 70% carbon and is used as a fuel in places where the smoke it produces is not a problem.

The first real coal is formed when the pressure has raised the "rank," or percentage of carbon, to 80%. This is **bituminous coal,** or "soft" coal. It is black, and does not break readily, but is still soft enough to rub off easily on one's hands. This is the most abundant variety of coal and is the type most commonly used in this country to fuel electric generating plants. In the past it was used to heat homes and to power the railroads. This coal is also the final product of purely sedimentary processing. Any further refinement requires a degree of metamorphism, brought about by an increase in

temperature as well as an increase in pressure. Obviously, very high temperatures would destroy the coal, but moderate temperatures serve to purify and harden it. Bituminous coal, when subjected to increased temperatures and pressure, metamorphoses into **anthracite.** Anthracite is "hard" coal, which is 90% carbon, black and glassy, and too hard to rub off on one's hands. Anthracite is difficult to ignite, but once lit, burns with a hot and clean flame.

All types of coal are found in layers, or **seams,** which vary in thickness from less than a meter to several tens of meters. Frequently, several seams will appear together, one over the next, and each will cover an area of many square kilometers. If our beliefs about the origin of coal are correct, each of these seams must represent the bottom of a large and ancient swamp. Furthermore, the thickness of the coal seams, and their frequent occurrence in the same location, suggest that the swamps and the conditions which produced them must have persisted for very long periods of time. At present, the best way to explain this is to associate the formation of coal deposits with major episodes of orogeny. As plates converge and mountain ranges grow, continental drainage is seriously obstructed. It is likely that this results in the formation of a series of large swamplands on the landward side of the emerging mountains. In the United States, this has led to significant accumulations of coal on the "interior" side of each of the major ranges of mountains. There are even some small accumulations in the far west associated with the Cascades.

Hydroelectric Power

At present, hydroelectric power is the only means by which solar energy is being harnessed. It is, of course, the sun's heat which evaporates water from the earth's surface. This water condenses in the atmosphere, falls back to the surface, and runs off in streams. It is easy to take advantage of this system by building a dam across one of the streams and forcing the water to turn an electric generator as it passes through. In one form or another, people have been using this type of power for thousands of years. In the past, the water turned mill wheels instead of generators, but the system has always had the same set of advantages and disadvantages. On the positive side, it is inexpensive and clean; on the negative side, it has always been limited by a lack of suitable locations for dams. Originally, locations were considered unsuitable because of their distance from centers of population or because of the engineering problems they presented; today, our judgment is more apt to be based on environmental concerns. Whatever the cause, a lack of dam sites is likely to confine this source of power to its present 4 or 5% share of this country's energy supply.

Nuclear Fission

A complete discussion of nuclear fission and its applications as a source of power is beyond the scope of this book. However, the fact that this source of energy is currently expected to expand, and the additional fact that some of its limitations are geological, make it important that we describe it briefly. The word "fission" means "breaking up," and therefore **nuclear fission** is the breaking up of atomic nuclei. As mentioned in Chapter 3, there are certain naturally occurring elements that break up spontaneously into other elements. In essence, this breakup is the result of the breaking apart of their nuclei into the nuclei of several smaller elements. Often, there are some atomic particles, and some energy, left over after the breakup, and these are given off as **radiation.** There are a few isotopes, or forms of elements, which are known as **fissionable isotopes** because they can be made to break up at an increased rate if we bombard their nuclei with neutrons. One such isotope is uranium 235, which occurs as 0.7% of naturally occurring uranium. In a **nuclear reactor,** this isotope is mixed with inert ingredients, which slow its rate of reaction, and tightly sealed in a series of tubes. These fuel elements are submerged in a fluid (usually water) and bombarded with neutrons. As the uranium nuclei break up into a number of radioactive products, a great deal of heat is generated. The radioactive products are trapped in the tubes, and the heat generated is transferred to the fluid and eventually used to produce steam. The steam, in turn, may be used to drive an electric generator, to power a submarine, or even to heat homes.

For years, nuclear power was touted as the "way of the future." It was claimed to be clean, safe and inexpensive. Experience has shown it to be none of these. Its problems appear to be twofold. In the first case, the highly radioactive by-products cannot be neutralized, and must therefore be permanently sealed off from the environment. At the site of a reactor, this is an engineering problem, but it eventually becomes a geological problem when we consider that we will need to find places in the earth where these "wastes" can be securely held for thousands of years. Dealing with these engineering and geological problems has become increasingly expensive.

Second, we have already mentioned that the fuel for these reactors occurs as a very small percentage of the known ores or uranium. This means that it is subject to depletion in about the same period of time as our current supplies of petroleum and coal. If an adequate substitute is not found, uranium will soon become very expensive.

Summary

This chapter deals with the ways in which the study of geology is applied to the search for economically valuable materials. These materials originate in the earth's crust, but the majority of them are rare. In order for us to be able to find them and cover the cost of extracting them from the ground, these materials must be concentrated into ores or trapped by a variety of geological processes. Most of the processes are associated with igneous activity and weathering. The more we know about these processes and the better we understand them, the more successful we are going to be.

Volcanism
    at convergent plate boundaries, 57
    at divergent plate boundaries, 36, 53
    at hot spots, 65–67
    on Mars, 114–115
    on moon, 113
    on Venus, 116
Volcanoes, 3

Water as agent of erosion, 71–73
Wave base, 106
Wavelength, of water waves, 106
Waves in water, 105–107
Weathering, 10, 69–70
    and formation of ores, 121
Wegener, Alfred, 1, 19, 20
Wilson cycle, 93
Wilson, Tuzo, 93
Wind
    erosion by, 72
    origin of, 109
Wind driven currents, 109–110
Winnowing, 76

Yellowstone Park, 67